Random Walks in Electrical Networks, Second Edition

Stefan Hollos and J. Richard Hollos

Random Walks in Electrical Networks, Second Edition
by Stefan Hollos and J. Richard Hollos

Abrazol Publishing
an imprint of Exstrom Laboratories LLC
662 Nelson Park Drive, Longmont, CO 80503-7674 U.S.A.

Publisher's Cataloging in Publication Data
Hollos, Stefan
Random Walks in Electrical Networks, Second Edition / by Stefan Hollos and J. Richard Hollos
p. cm.
ISBN: 978-1-887187-49-7
Library of Congress Control Number: 2025948834
1. Electric networks. 2. Random walks (Mathematics). 3. Electric network topology.
I. Title. II. Hollos, Stefan.
TK454.2 .H65567 2026
621.319 HOL

Cover image: Camille Pissarro - The Garden of the Tuileries on a Winter Afternoon, 1899, https://upload.wikimedia.org/wikipedia/commons/0/00/Camille_Pissarro%2C_The_Garden_of_the_Tuileries_on_a_Winter_Afternoon%2C_1899.jpg

Contents

1

PREFACE

To show that the random walk method can be profitably applied in almost every area of circuit analysis we have greatly expanded the number of applications covered in this new edition. Six new chapters have been added to the book which almost doubles the size of the original.

We start with circuit transformations. In particular, we look at the delta-wye transformation and the more general star-mesh transformation.

Next we show how to apply the method to circuits containing operational amplifiers. We cover the inverting, non-inverting and difference amplifier as well as the integrator and active filters.

Some of the more exotic op amp applications covered are the negative impedance converter and the gyrator. The software available on the book's website is useful for dealing with some of the larger and more complex op amp circuits.

We end the new applications with a look at discrete transistor circuits. We only scratch the surface here. A motivated reader could have fun extending this to circuits not covered in the book. If you do, please let us know. Intellectual adventures are best when shared. Also be sure to check the book's website at
https://www.abrazol.com/books/rwnet2/
for software updates and further developments.

Stefan Hollos
Richard Hollos
Exstrom Laboratories LLC
Longmont, Colorado
December 2025

Preface to First Edition

Electrical network analysis is an old, mature and seemingly closed subject. Georg Ohm (16 March 1789 – 6 July 1854) published his law in 1827, almost 200 years ago and Gustav Kirchhoff's (12 March 1824 – 17 October 1887) circuit laws date to 1845. One of the more recent advances in network theory is Tellegen's theorem, published by Bernard Tellegen (24 June 1900 – 30 August 1990) in 1952.

But network analysis is far from being a closed subject. It is part of the much broader subject of general network theory which is itself a branch of the mathematical subject of graph theory. This means that many of the powerful methods of graph theory can be applied to electrical network analysis. Once an electrical network has been reduced to a graph we can use the topology, symmetry, and the walk structure of the graph to solve problems related to the underlying electrical network.

Much of the work in this area has not made its way out of the confines of academic research. The electrical engineer is unlikely to be exposed to this way of doing network analysis in their undergraduate education. They may come across some of it in graduate school or later in their career but the practical applications of it may not always be clear. It is our belief that this way of doing network analysis is exciting, interesting and has practical benefits that have yet to be fully realized.

This book is our attempt to bring some of these nonstandard methods of network analysis to the attention of more engineers. In particular, we will show in this book how some of the traditional methods of doing network analysis can be turned into an equivalent problem involving a random walk on a graph. We will show that this probabilistic interpretation of electrical networks leads to graphical techniques that can make it easier to solve circuit analysis problems. The book contains a variety of problems that are solved in detail using this method. We have tried not to make this an academic book. It should be of interest to any engineer looking for new and more efficient ways to do network analysis. The only prerequisite for this book is a good knowledge of standard electrical network theory.

Finally, the website for the book you are now reading is
https://www.abrazol.com/books/rwnet/
where we will post related resources.

Stefan Hollos
Richard Hollos
Exstrom Laboratories LLC
Longmont, Colorado
December 2023

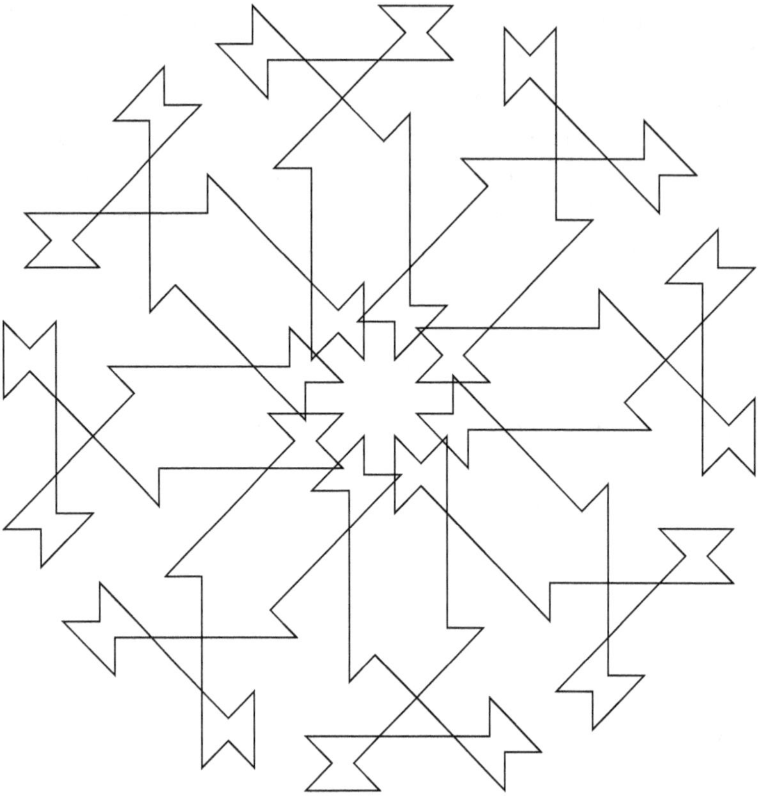

No wine for me. Strange enough things happen when my head is clear. I want to know the difference.

Robert Jordan, The Wheel of Time

2

NODAL ANALYSIS WITH A VOLTAGE SOURCE

When a voltage source is connected across two nodes of a network it will induce voltages at the other nodes. Those other voltages can be found using a method called nodal analysis. We will show how this method works, how it can be used to determine the equivalent admittance or impedance of the network and how it can be turned into a corresponding problem involving a random walk on a graph.

Let's call the number of nodes in the network $n+2$. A network must have a minimum of 2 nodes, in which case $n = 0$. Label the nodes from 0 to $n+1$ with the voltage source connected between nodes 0 and $n+1$ such that node $n+1$ is the ground node (at zero potential). The nodes to which the source is not connected are labeled from 1 to n.

Suppose node 0 (the source node) is connected to m other nodes through admittances $Y_{01}, Y_{02}, \ldots, Y_{0m}$ as shown in figure 2.1. There will be other connections to node $n+1$ (ground) which are not shown in the figure or else this would be a dead network.

We can find an expression for the equivalent admittance of the network by applying Kirchhoff's current law to node 0. If V_i is the voltage at node i then the current that flows from node 0 to i is given by $(V_0 - V_i)Y_{0i}$. If I is the total current flowing into node 0 from the voltage source, then

Kirchhoff's current law tells us that

$$\sum_{i=1}^{m}(V_0 - V_i)Y_{0i} = I \tag{2.1}$$

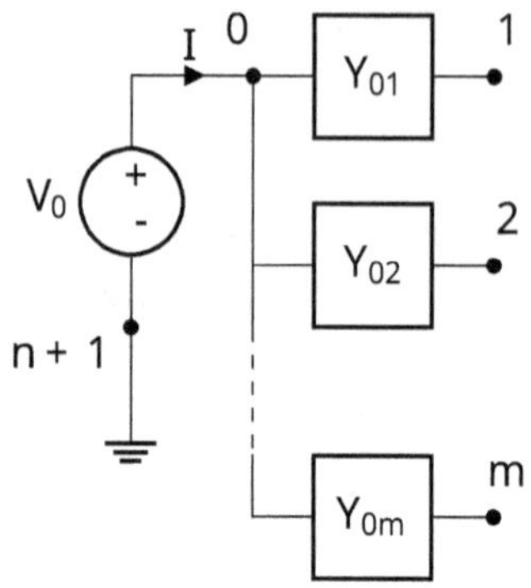

Figure 2.1: Voltage source connected to a network.

We can get an expression for the equivalent admittance of the network by simply dividing both sides of equation 2.1 by V_0. This gives us the equivalent admittance connected across nodes 0 and $n+1$ of the network. Calling this admittance Y_0, we have

$$Y_0 = \frac{I}{V_0} = \sum_{i=1}^{m}(1 - g_i)Y_{0i} \tag{2.2}$$

where $g_i = V_i/V_0$ is a voltage transfer function from the source to node i.

To calculate Y_0 we still need to find the g_i values. We can find them by applying Kirchhoff's current law to the remaining nodes, 1 thru n, of the network (the ground node is not included). This will result in a system of n equations that can then be solved for all the g_i, not just the first m required in equation 2.2.

In general for node k, assuming there are no other sources connected to the network, Kirchhoff's current law says that

$$\sum_i (V_k - V_i)Y_{ki} = 0 \tag{2.3}$$

where the sum is over all nodes connected to node k. The equation simply says that, since there is no source connected to the node, the sum of all currents leaving the node must equal 0. Collecting the V_k terms, we can rewrite equation 2.3 as follows

$$V_k \sum_i Y_{ki} - \sum_i V_i Y_{ki} = 0 \tag{2.4}$$

For nodes $k = 1, 2, \ldots, m$, one of the nodes connected to k will be node 0 (the source node). If we move that term to the right hand side then equation 2.4 becomes

$$V_k \sum_i Y_{ki} - \sum_{i \neq 0} V_i Y_{ki} = V_0 Y_{k0} \tag{2.5}$$

Now divide both sides of equations 2.4 and 2.5 by

$$V_0 \sum_j Y_{kj} \tag{2.6}$$

where the sum is over all nodes connected to node k. If we let $g_k = V_k/V_0$ be a voltage transfer function and

$$p_{ki} = \frac{Y_{ki}}{\sum_j Y_{kj}} \tag{2.7}$$

then equation 2.4, for the nodes that are not connected to node 0, becomes

$$g_k - \sum_i g_i p_{ki} = 0 \tag{2.8}$$

and equation 2.5, for the nodes that are connected to node 0, becomes

$$g_k - \sum_{i \neq 0} g_i p_{ki} = p_{k0} \tag{2.9}$$

Note that the sum in equation 2.8 is over all nodes connected to node k and the sum in equation 2.9 is over all nodes connected to node k except for node 0.

We will call the p_{ki} values transition probabilities because in most cases they behave exactly like probabilities. When the Y_{ki} are real, i.e. conductances, the p_{ki} will be real values less than 1. When the Y_{ki} are complex the p_{ki} will in general not be real but will have magnitudes less than 1. In both cases the p_{ki} are clearly analogous to probabilities in the sense that the sum of all p_{ki} for node k will equal 1. Note that when we say all the p_{ki} that includes the p_{ki} for connections to the ground node which were not used in the formulation of the node equations given above. When connections to the ground node are included we have

$$\sum_i p_{ki} = 1 \tag{2.10}$$

When we formulate the network problem as a random walk problem we will see that p_{ki} is the probability of jumping from node k to node i.

Before looking at the random walk formulation let's look at the more traditional method of solving the problem. It is clear that equations 2.8 and 2.9 must be solved for the voltage transfer functions $g_i = V_i/V_0$ which then determine the node voltages V_i given the source voltage V_0. The g_i values for $i = 1$ through m can then be used to determine the equivalent admittance of the network via equation 2.2.

One way to proceed with such a solution is to put equations 2.8 and 2.9 into matrix form as follows

$$(I - A)\vec{g} = \vec{p} \tag{2.11}$$

where $\vec{g} = [g_1\ g_2 \cdots g_n]^T$ is the vector of voltage transfer functions, $\vec{p} = [p_{10} \cdots p_{m0}\ 0 \cdots 0]^T$ is the vector of transition probabilities between node 0 and the m nodes connected to it, and A is the transition probability matrix for the network node graph.

In the current context, the network node graph is a graph where the vertices correspond to nodes 1 to n of the network. Vertex i is connected to vertex j in the graph if node i is connected to node j in the network. The connections are weighted by the transition probabilities and the n by n matrix A tabulates these connections. The element in row i and column j of the matrix A will equal the transition probability p_{ij} that connects node i to node j. (From here on we will refer to both the vertices of the graph and the nodes of the network as nodes)

Solving equation 2.11 for $\vec{g}$ involves inverting the matrix $I - A$. The solution is then

$$\vec{g} = (I - A)^{-1}\vec{p} \tag{2.12}$$

This is a purely abstract formulation of the problem and its solution. We will now analyze a concrete problem in detail which should lead to better understanding of the formalism and how to use it.

Figure 2.2: Michael Faraday (1791-1867) did much of the experimental research that forms the basis for understanding how circuit elements behave. Public domain image courtesy of Wikimedia Commons at https://commons.wikimedia.org/wiki/File:Michael_Faraday_sitting_crop.jpg

3

RESISTIVE BRIDGE CIRCUIT

The first network we will analyze is shown in figure 3.1. It is a simple resistive bridge network with 4 nodes and three unique resistance values: a, b, c. A voltage source is connected between nodes 0 and 3. The problem is to find the voltages at nodes 1 and 2 and the equivalent resistance seen by the voltage source.

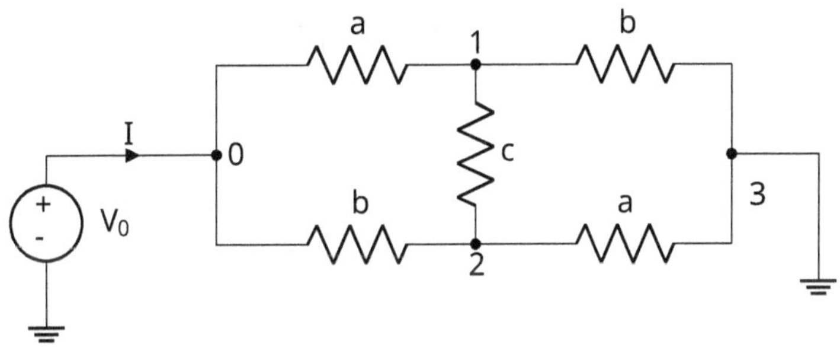

Figure 3.1: Resistive bridge circuit.

If we sum the currents at node 1 we get

$$\frac{V_1 - V_0}{a} + \frac{V_1 \quad V_2}{c} + \frac{V_1}{b} = 0 \tag{3.1}$$

Collecting terms, we can write this as

$$\left(\frac{1}{a} + \frac{1}{b} + \frac{1}{c}\right) V_1 - \frac{1}{c} V_2 = \frac{1}{a} V_0 \tag{3.2}$$

Now summing the currents at node 2 we get

$$\frac{V_2 - V_0}{b} + \frac{V_2 - V_1}{c} + \frac{V_2}{a} = 0 \tag{3.3}$$

Collecting terms, we can write this as

$$-\frac{1}{c}V_1 + \left(\frac{1}{a} + \frac{1}{b} + \frac{1}{c}\right) V_2 = \frac{1}{b}V_0 \tag{3.4}$$

If we divide equations 3.2 and 3.4 by

$$\left(\frac{1}{a} + \frac{1}{b} + \frac{1}{c}\right) V_0 \tag{3.5}$$

then we get the following system of equations to solve for g_1 and g_2

$$\begin{aligned} g_1 - pg_2 &= p_{10} \\ -pg_1 + g_2 &= p_{20} \end{aligned} \tag{3.6}$$

where $g_1 = V_1/V_0$, $g_2 = V_2/V_0$, and

$$p = p_{12} = p_{21} = \frac{1/c}{1/a + 1/b + 1/c} \tag{3.7}$$

is the transition probability from node 1 to 2 or 2 to 1. On the right hand side we have the transition probabilities from nodes 1 and 2 to the source, node 0.

$$p_{10} = \frac{1/a}{1/a + 1/b + 1/c} \tag{3.8}$$

$$p_{20} = \frac{1/b}{1/a + 1/b + 1/c} \tag{3.9}$$

In matrix form equation 3.6 is

$$\begin{bmatrix} 1 & -p \\ -p & 1 \end{bmatrix} \begin{bmatrix} g_1 \\ g_2 \end{bmatrix} = \begin{bmatrix} p_{10} \\ p_{20} \end{bmatrix} \tag{3.10}$$

The network node graph for this problem is rather simple, consisting of only nodes 1 and 2 connected to each other. Comparing equation 3.10 with equation 2.11 we see that the transition probability matrix is

$$A = \begin{bmatrix} 0 & p \\ p & 0 \end{bmatrix} \tag{3.11}$$

This matrix says that, leaving out the source and ground nodes, the network only consists of two nodes, 1 and 2, connected to each other with transition probability p.

The matrix $(I - A)$ in equation 3.10 can be easily inverted to give the following solution

$$\begin{bmatrix} g_1 \\ g_2 \end{bmatrix} = \frac{1}{1-p^2} \begin{bmatrix} 1 & p \\ p & 1 \end{bmatrix} \begin{bmatrix} p_{10} \\ p_{20} \end{bmatrix} \tag{3.12}$$

So we have

$$g_1 = \frac{p_{10} + p\, p_{20}}{1-p^2} \tag{3.13}$$

$$g_2 = \frac{p\, p_{10} + p_{20}}{1-p^2} \tag{3.14}$$

These are the voltage transfer functions $g_1 = V_1/V_0$ and $g_2 = V_2/V_0$. Given V_0 they give us the voltages V_1 and V_2. Using the expressions in equations 3.7, 3.8 and 3.9 we have

$$g_1 = \frac{(a+c)b}{(a+b)c + 2ab} \tag{3.15}$$

$$g_2 = \frac{(b+c)a}{(a+b)c + 2ab} \tag{3.16}$$

and when $V_0 = 1$ the voltage across the resistance c is

$$g_1 - g_2 = \frac{(b-a)c}{(a+b)c + 2ab} \tag{3.17}$$

We can see that when $a = b$ there is no voltage across the resistor and $g_1 = g_2 = 1/2$ as we would expect.

The g_1 and g_2 values can now be used to find the equivalent resistance of the network. If we sum the currents at node 0 we get

$$\frac{V_0 - V_1}{a} + \frac{V_0 - V_2}{b} = I \tag{3.18}$$

where I is the current flowing into the network from the voltage source. Divide both sides of the equation by V_0 to get an expression for the equivalent resistance, R_0.

$$\frac{1}{R_0} = \frac{I}{V_0} = \frac{1-g_1}{a} + \frac{1-g_2}{b} \tag{3.19}$$

Using the values for g_1 and g_2 given above we get

$$R_0 = \frac{(a+b)c + 2ab}{a + b + 2c} \tag{3.20}$$

4

A RANDOM WALK INTERPRETATION

To see how the resistive bridge circuit problem corresponds to a random walk on a graph consider the graph shown in figure 4.1. The nodes in the graph mirror the nodes in the circuit in figure 3.1. All nodes in the circuit are included in the graph except for the ground node. Transition probabilities from nodes 1 to 2 and from 2 to 1 are both equal to p. Transition probabilities from 1 to 0 and 2 to 0 are p_{10} and p_{20} respectively. Note that there are no transition probabilities from 0 to 1 or 2. When a random walk reaches 0 it stops.

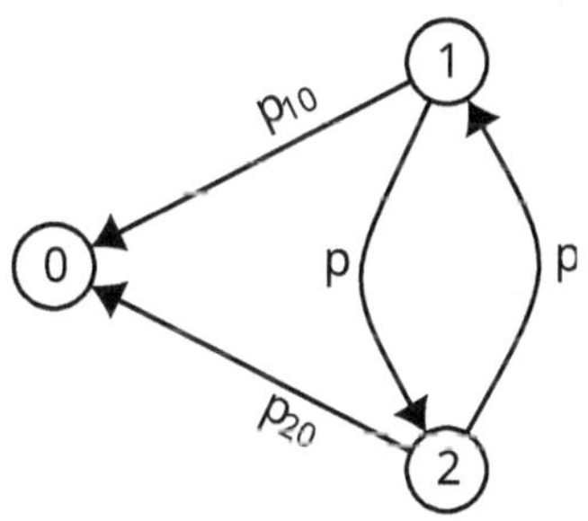

Figure 4.1: Bridge circuit graph.

What is the probability that a random walk starting at node 1 ends at node 0? There are two ways to get from 1 to 0. In the first way there will be zero or more round trips from 1 to 2 and back followed by a final step to 0. Each round trip has probability p^2 so the total probability for this way is

$$p_{10}(1 + p^2 + p^4 + p^6 \cdots) = \frac{p_{10}}{1 - p^2} \tag{4.1}$$

In the second way there will also be zero or more round trips from 1 to 2 and back followed one more step to 2 and a final step to 0. The total probability for this way is

$$p\, p_{20}(1 + p^2 + p^4 + p^6 \cdots) = \frac{p\, p_{20}}{1 - p^2} \tag{4.2}$$

The probability of starting at 1 and ending at 0 will then be the sum of these last two equations

$$\frac{p_{10} + p\, p_{20}}{1 - p^2} = g_1 \tag{4.3}$$

In the same way, we can show that the probability that a random walk starting at node 2 ends at 0 is given by

$$\frac{p_{20} + p\, p_{10}}{1 - p^2} = g_2 \tag{4.4}$$

The probabilities g_1 and g_2 in equations 4.3 and 4.4 are the same as the voltage transfer functions g_1 and g_2 in equations 3.13 and 3.14. If you define the transition probabilities as in equations 3.7, 3.8 and 3.9 then the network problem is identical to this random walk problem.

The general procedure for turning any network problem with a voltage source into a random walk problem is to create a graph for the network that contains all but the ground node. The transition probability going from node i to j in the graph is given by

$$p_{ij} = \frac{Y_{ij}}{\sum_k Y_{ik}} \tag{4.5}$$

where Y_{ij} is the admittance between i and j and the sum is over all nodes connected to i. There are no transitions out of node 0 (the source node).

The voltage transfer function $g_i = V_i/V_0$ is identical to the probability of starting at node i in the graph and ending at node 0. One way to solve for g_i is to use the traditional method of constructing the transition probability matrix A that appears in equation 2.11 where element $A_{i,j}$ of the matrix is equal to p_{ij}. The matrix $I - A$ is inverted to find the vector of voltage transfer functions as in equation 2.12. The probabilistic interpretation of g_i can be seen from the expansion

$$(I - A)^{-1} = I + A + A^2 + A^3 + \cdots \tag{4.6}$$

Since element $A_{i,j}^{(n)}$ of A^n is the probability of a walk starting at i and ending at j after n steps the meaning of equation 2.12 becomes clear. It says that g_i is

$$g_i = \sum_j \begin{pmatrix} \text{sum of probabilities of} \\ \text{all walks from i to j} \end{pmatrix} \cdot (\text{probability of stepping from j to 0}) \tag{4.7}$$

where the sum is over all the nodes j that are connected to 0.

But matrix inversion is not the only way to calculate $\vec{g}$. Individual g_i can be calculated using graphical techniques. A simple example of this was shown in the above solution for the resistive bridge network. This is where the real power of turning the network problem into a random walk on a graph becomes apparent. These graphical solution techniques will be demonstrated in more detail in the following examples.

Figure 4.2: James Clerk Maxwell (1831-1879) along with Michael Faraday discovered the fundamental physical laws upon which the behavior of electrical circuit elements is based. Public domain image courtesy of Wikimedia Commons at https://commons.wikimedia.org/wiki/File:James_Clerk_Maxwell.png

5

EQUAL RESISTANCE LADDERS

The equal resistance ladder offers one of the most straightforward examples of using the random walk method. The simplest equal resistance ladder is the four resistor network shown in figure 5.1. The network has four nodes and all resistors have the same value R. A voltage source is connected between nodes 0 and 3. We want to find the voltages at nodes 1 and 2.

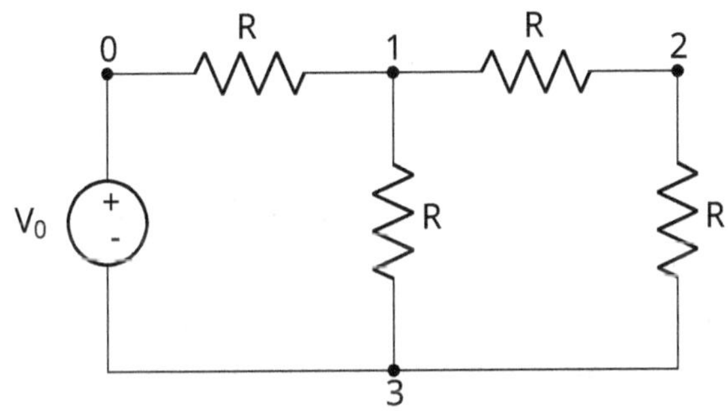

Figure 5.1: 2 rung resistance ladder.

The graph for this network is shown in figure 5.2. Since node 1 of the network has three equal resistors connected to it, all transition probabilities out of this node in the graph are equal to 1/3. Since node 2 of the network has two equal resistors connected to it, all transition probabilities out of this node in the graph are equal to 1/2. So we have $p_{10} = p_{12} = 1/3$ and $p_{21} = 1/2$.

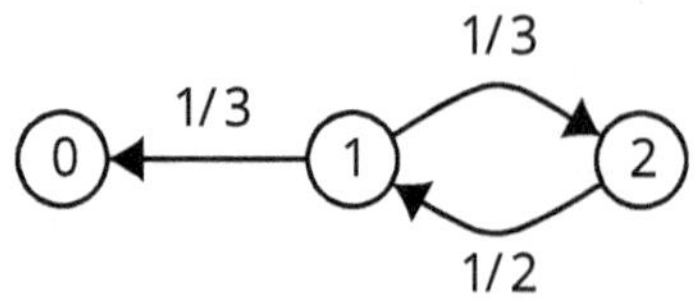

Figure 5.2: 2 rung ladder graph.

Equation 2.11 for both the network and graph is then

$$\begin{bmatrix} 1 & -1/3 \\ -1/2 & 1 \end{bmatrix} \begin{bmatrix} g_1 \\ g_2 \end{bmatrix} = \begin{bmatrix} 1/3 \\ 0 \end{bmatrix} \tag{5.1}$$

and the solution using matrix inversion is

$$\begin{bmatrix} g_1 \\ g_2 \end{bmatrix} = \frac{6}{5} \begin{bmatrix} 1 & 1/3 \\ 1/2 & 1 \end{bmatrix} \begin{bmatrix} 1/3 \\ 0 \end{bmatrix} \tag{5.2}$$

Multiplying this out, we get $g_1 = 2/5$ and $g_2 = 1/5$.

Now we will show another way of arriving at this result using the graphical technique of node elimination. This involves removing a node from the graph and adjusting the connections of the nodes it was connected to so that transition probabilities remain consistent with the original graph. The new transition probabilities are two step probabilities that go through the deleted node. The procedure is best understood with some examples.

We will begin by eliminating node 2 in the graph. Since it is only connected to node 1, it can be eliminated by putting in a loop back transition for node 1 that has probability $1/3 \cdot 1/2 = 1/6$ as shown in figure 5.3. This loop back represents the transition from 1 to 2 and back.

In a walk the loop back transition can be executed any number of times. Therefore, its contribution to the walk probability will be a geometric series in the probability of a single loop. If the probability of a single loop is p

then the total contribution of the loop is

$$1 + p + p^2 + p^3 + \cdots = \frac{1}{1-p} \tag{5.3}$$

where the p^n term in the series represents n loop transitions.

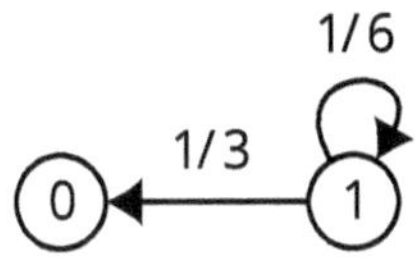

Figure 5.3: Elimination of node 2 in figure 5.2.

So from figure 5.3 we see that the probability of a walk starting at 1 and ending at 0 is the loop back transition times the transition probability from 1 to 0 or

$$g_1 = \frac{1/3}{1-1/6} = \frac{2}{5} \tag{5.4}$$

This is the same result we got above using matrix inversion.

We could now get the probability of starting at 2 and ending at 0 by eliminating node 1 but there is a simpler way. The probability can be found by noting that any walk from 2 to 0 must begin with a step from 2 to 1 with probability 1/2. From there we have a walk that starts at 1 and ends at 0 but we already know that probability. It's g_1 which we calculated above. So we have

$$g_2 = \frac{1}{2} g_1 = \frac{1}{5} \tag{5.5}$$

Which is again the same result found by matrix inversion.

In this particular case we could also find g_2 by translating back to the language of voltages. In terms of voltages we have

$$g_2 = \frac{V_2}{V_0} = \frac{V_2}{V_1}\frac{V_1}{V_0} = \frac{1}{2} g_1 \tag{5.6}$$

where we have used the fact that $V_1/V_0 = g_1$ and from the network diagram in figure 5.1 it is clear that we have a simple voltage divider where $V_2/V_1 = 1/2$.

The graphical technique of node elimination that we used here is very powerful. It can be used to calculate some or all of the g_i values without the need to invert a matrix. There will be more examples of how it is used in the following problems.

Let's now look at the slightly larger resistance ladder shown in figure 5.4. The network has five nodes and once again all resistors have the same value R. A voltage source is connected between nodes 0 and 4. We want to find the voltages at nodes 1, 2 and 3.

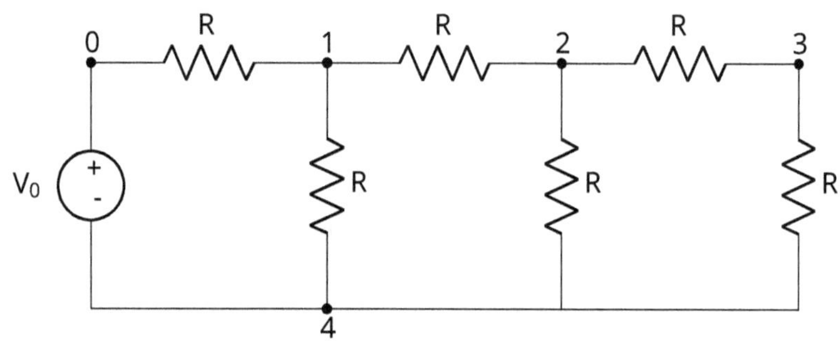

Figure 5.4: 3 rung resistance ladder.

The graph for this network is shown in figure 5.5. Nodes 1 and 2 of the network both have three equal resistors connected to them so all transition probabilities out of these nodes in the graph are equal to 1/3. Node 3 of the network has two equal resistors connected to it so all transition probabilities out of this node in the graph are equal to 1/2. The graph transition probabilities are then $p_{10} = p_{12} = p_{21} = p_{23} = 1/3$ and $p_{32} = 1/2$.

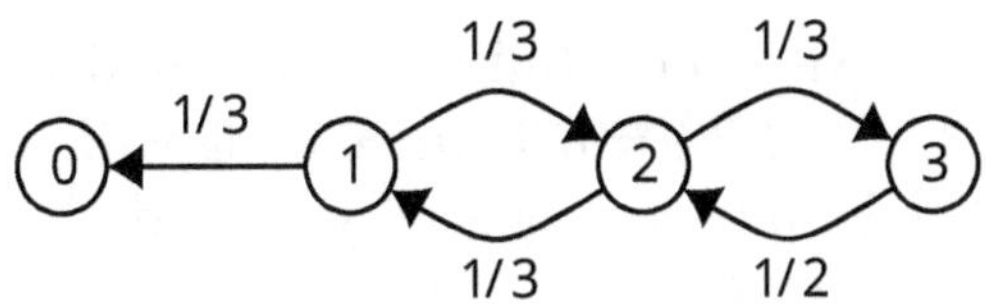

Figure 5.5: 3 rung ladder graph.

With these probabilities we can construct the following matrix equation

$$\begin{bmatrix} 1 & -1/3 & 0 \\ -1/3 & 1 & -1/3 \\ 0 & -1/2 & 1 \end{bmatrix} \begin{bmatrix} g_1 \\ g_2 \\ g_3 \end{bmatrix} = \begin{bmatrix} 1/3 \\ 0 \\ 0 \end{bmatrix} \tag{5.7}$$

and the solution using matrix inversion is

$$\begin{bmatrix} g_1 \\ g_2 \\ g_3 \end{bmatrix} = \frac{1}{13} \begin{bmatrix} 5 & 16 & 2 \\ 6 & 18 & 6 \\ 3 & 9 & 16 \end{bmatrix} \begin{bmatrix} 1/3 \\ 0 \\ 0 \end{bmatrix} \tag{5.8}$$

Multiplying this out, we get $g_1 = 5/13$, $g_2 = 2/13$ and $g_3 = 1/13$.

To solve this problem using the graphical method we start by eliminating node 3. This is done by putting a loop back on node 2 as shown in figure 5.6. The loop back has probability $1/3 \cdot 1/2 = 1/6$.

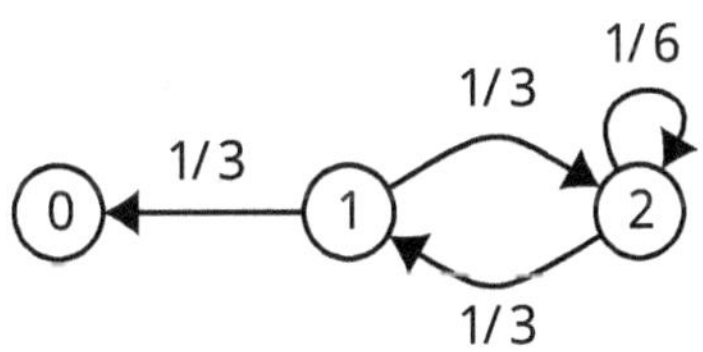

Figure 5.6: Elimination of node 3 in figure 5.5.

Next we eliminate node 2. This is done by putting a loop back on node 1 as shown in figure 5.7. To find the probability of this loop back, note that it involves a step from 1 to 2 with probability 1/3 followed by the loop back on 2 which has weight $1/(1-1/6) = 6/5$ and ending with a step back to 1 with probability 1/3. The loop back probability on node 1 is therefore $1/3 \cdot 6/5 \cdot 1/3 = 2/15$.

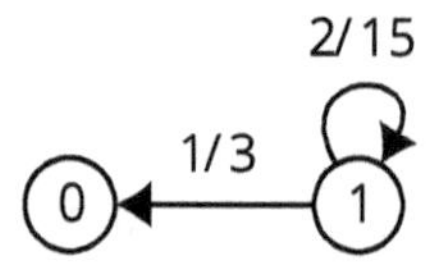

Figure 5.7: Elimination of node 2 in figure 5.6.

From figure 5.7 we see that g_1 is given by the loop back on node 1 followed by the transition to 0 with probability 1/3.

$$g_1 = \frac{1/3}{1-2/15} = \frac{5}{13} \tag{5.9}$$

To calculate g_2, note that it is the probability of moving from 2 to 1 times g_1. From figure 5.6 we see that the probability of moving from 2 to 1 is given by loop back on 2 followed by the probability of stepping from 2 to 1. So we have

$$g_2 = \frac{1/3}{1-1/6}\, g_1 = \frac{2}{5} \cdot \frac{5}{13} = \frac{2}{13} \tag{5.10}$$

Finally, g_3 is the probability of stepping from 3 to 2 times g_2. So we have $g_3 = 1/2 \cdot g_2 = 1/13$. These are the same values that we got through matrix inversion but they were found by looking at walks on a graph where the connections between nodes are weighted with probabilities.

The way to extend this to longer ladders should be now be clear. As an exercise try solving for the voltages of a four rung ladder. Show that the four g_i values are: $g_1 = 13/34$, $g_2 = 5/34$, $g_3 = 2/34$, $g_4 = 1/34$. Show that for a five rung ladder $g_1 = 34/89$, $g_2 = 13/89$, $g_3 = 5/89$, $g_4 = 2/89$, $g_5 = 1/89$. Do you see a pattern? Can you predict the values for the six rung ladder or the n rung ladder?

6

RESISTANCE LATTICES

A very simple resistance lattice is shown in figure 6.1. The network has five nodes with all resistors equal except for the two connected from node 1 to 4 and from node 3 to 4. A voltage source is connected between nodes 0 and 4, making 4 the ground node. We want to find the voltages at nodes 1, 2 and 3 as well as the equivalent resistance of the network.

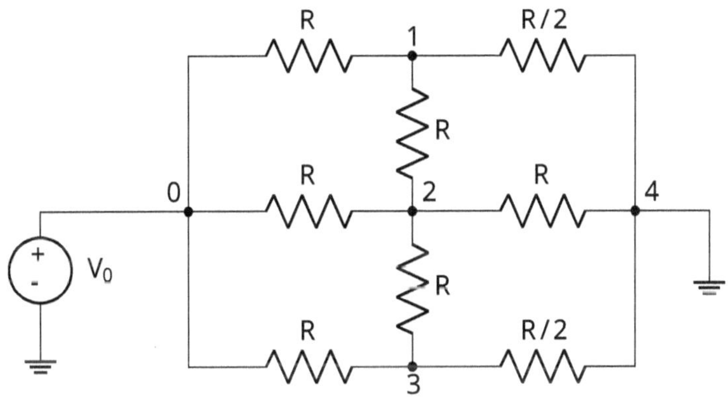

Figure 6.1: Resistor lattice network.

The graph for this network is shown in figure 6.2. All the transition probabilities in the graph are equal to $p = 1/4$. The transition probability matrix is then given by

$$A = \begin{bmatrix} 0 & p & 0 \\ p & 0 & p \\ 0 & p & 0 \end{bmatrix} \tag{6.1}$$

Calculating the inverse $(I - A)^{-1}$ we get the following solution for the g_i values.

$$\begin{bmatrix} g_1 \\ g_2 \\ g_3 \end{bmatrix} = \frac{1}{1-2p^2} \begin{bmatrix} 1-p^2 & p & p^2 \\ p & 1 & p \\ p^2 & p & 1-p^2 \end{bmatrix} \begin{bmatrix} p \\ p \\ p \end{bmatrix} \tag{6.2}$$

Multiplying this out, we get

$$g_1 = g_3 = \frac{p(1+p)}{1-2p^2} \tag{6.3}$$

$$g_2 = \frac{p(1+2p)}{1-2p^2} \tag{6.4}$$

Using the value $p = 1/4$ we get $g_1 = g_3 = 5/14$, $g_2 = 3/7$.

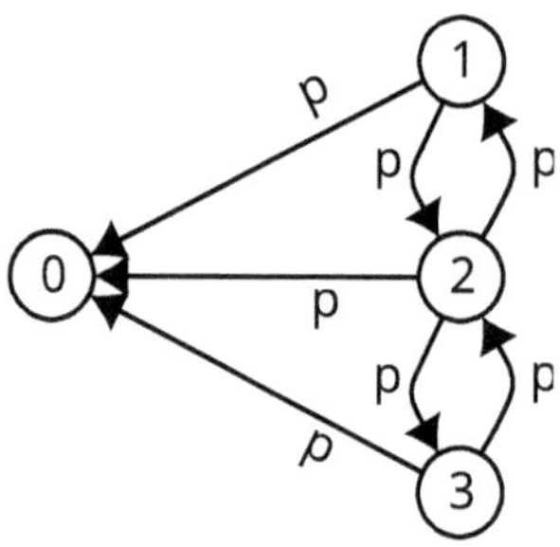

Figure 6.2: Graph for the network in figure 6.1.

Now let's solve the problem using graphical techniques. From the symmetry of the graph it is clear that $g_1 = g_3$ so we will begin by eliminating nodes 1 and 3 to solve for g_2.

The loop back probability going from 2 to 1 and back is p^2. This is the same as the loop back probability going from 2 to 3 and back. The total loop back probability on 2, from eliminating 1 and 3, is therefore $2p^2$, as shown in figure 6.3.

With nodes 1 and 3 eliminated the transition probability from 2 to 0 must now include going through 1 or 3 which has probability $2p^2$ so the total transition probability going from 2 to 0 is $p + 2p^2$.

The probability of starting at 2 and ending at 0 is therefore

$$g_2 = \frac{p + 2p^2}{1 - 2p^2} = \frac{p(1 + 2p)}{1 - 2p^2} \tag{6.5}$$

This matches equation 6.4, calculated by matrix inversion.

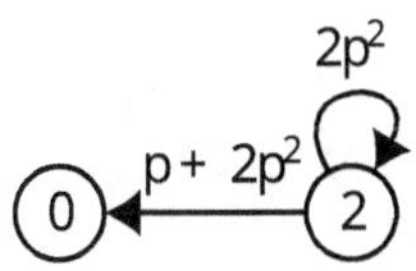

Figure 6.3: Graph in figure 6.2 with nodes 1 and 3 eliminated.

Now we want to calculate g_1, which is the probability of starting at 1 and ending at 0. We could solve for it using node elimination but we can also get the answer from the value of g_2 we just calculated. To start at 1 and end at 0 we can directly transition from 1 to 0 with probability p or we can transition to 2 with probability p and from there end up at 0. This means that $g_1 = p + pg_2$. Substituting the value of g_2 from equation 6.5 we get

$$g_1 = p + \frac{p^2(1 + 2p)}{1 - 2p^2} = \frac{p(1 + p)}{1 - 2p^2} \tag{6.6}$$

This matches equation 6.3, calculated by matrix inversion.

This problem illustrates the fact that, due to the symmetry of the graph, eliminating some nodes first makes the problem easier to solve. To see this, try solving the problem again by first eliminating node 2.

Figure 6.4: Gustav Robert Kirchhoff (1824-1887) is a major contributor to the fundamental understanding of electrical circuits. Public domain image courtesy of Wikimedia Commons at https://commons.wikimedia.org/wiki/File:Gustav_Robert_Kirchhoff.jpg

7

NODAL ANALYSIS WITH A CURRENT SOURCE

Another way to excite a network is with a current source as shown in figure 7.1. In this chapter we will show that such a network can also be analyzed using the random walk method.

As with the voltage source excitation, we will call the number of nodes in the network $n + 2$. The current source is connected so that current flows into node 0 and we assume that there are m branches coming out of 0 that connect to nodes labeled 1 thru m.

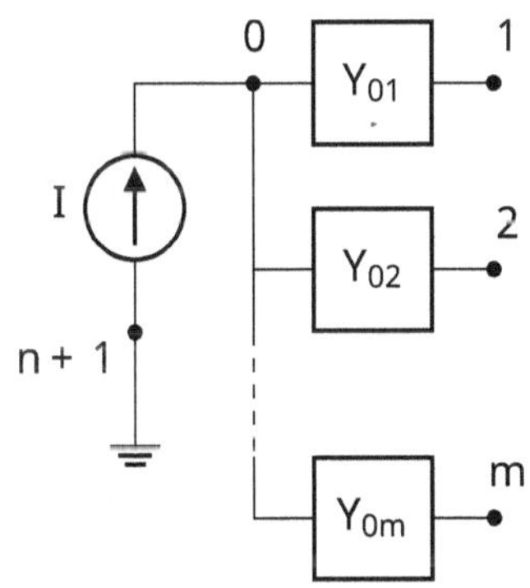

Figure 7.1: Current source connected to a network.

Applying Kirchhoff's current law to node 0, we get

$$\sum_{i=1}^{m}(V_0 - V_i)Y_{0i} = I \tag{7.1}$$

This is identical to equation 2.1. The only difference is that now I is the independent variable and all the V_i are dependent variables, including V_0. The equation can be written as

$$V_0 \sum_{i=1}^{m} Y_{0i} - \sum_{i=1}^{m} V_i Y_{0i} = I \tag{7.2}$$

We divide both sides of the equation by $\sum_i Y_{0i}$ and write it as follows

$$V_0 - \sum_{i=1}^{m} V_i p_{0i} = \frac{I}{\sum_i Y_{0i}} \tag{7.3}$$

where p_{0i} is the transition probability going from node 0 to i, defined as

$$p_{0i} = \frac{Y_{0i}}{\sum_j Y_{0j}} \tag{7.4}$$

For the nodes $k \neq 0$ we have

$$V_k \sum_i Y_{ki} - \sum_i V_i Y_{ki} = 0 \tag{7.5}$$

where the sums are over all branches connected to k. Divide both sides of the equation by $\sum_i Y_{ki}$ and write it as follows

$$V_k - \sum_i V_i p_{ki} = 0 \tag{7.6}$$

where p_{ki} is the transition probability going from node k to i, defined as

$$p_{ki} = \frac{Y_{ki}}{\sum_j Y_{kj}} \tag{7.7}$$

So there will be $n+1$ equations in the unknowns $V_0, V_1, \cdots, V_n$. In matrix form the system of equations is written as

$$(I - A)\vec{V} = \vec{b} \tag{7.8}$$

where $\vec{V} = [V_0\ V_1 \cdots V_n]^T$, $\vec{b} = [I/\sum_i Y_{0i}\ 0 \cdots 0]^T$ and A is the transition probability matrix for the network node graph that includes all but the ground node, i.e. nodes 0 through n. The element in row k and column i of A is equal to p_{ki} as given by equation 7.7.

The solution for $\vec{V}$ is then given, at least formally, by

$$\vec{V} = (I - A)^{-1}\vec{b} \tag{7.9}$$

If we let $B = (I - A)^{-1}$ then since only the first component of the vector $\vec{b}$ is nonzero it is clear that the multiplication $B\vec{b}$ in equation 7.9 results in

$$V_i = \frac{B_{i0} I}{\sum_j Y_{0j}} \tag{7.10}$$

keeping in mind that the admittance summation in the denominator is over all network branches connected to node 0.

In terms of random walks B_{i0} is the sum of the probabilities for all random walks that start at node i of the network graph and end at node 0. The sum includes walks of all possible lengths. This probabilistic interpretation for B_{i0} becomes clear when we note that $B = (I - A)^{-1}$ can be expanded as follows

$$B = (I - A)^{-1} = I + A + A^2 + A^3 + \cdots \tag{7.11}$$

The element in row i and column j of A^n is the probability that a walk on the network graph starting at node i will end at node j after exactly n steps. So B_{ij} is the sum of the probabilities for all walks of any length that start at i and end at j.

Equation 7.10 allows us to write the equivalent impedance of the network across nodes 0 and $n + 1$ as follows

$$Z_0 = \frac{V_0}{I} = \frac{B_{00}}{\sum_j Y_{0j}} \tag{7.12}$$

Figure 7.2: Georg Simon Ohm (1789-1854) discovered that the voltage across a conductor is proportional to the current through it. Public domain image courtesy of Wikimedia Commons at https://commons.wikimedia.org/wiki/File:Georg_Simon_Ohm_(1789-1854).jpg

8

COMPARISON OF VOLTAGE AND CURRENT SOURCE ANALYSIS

We have now covered two ways to measure or calculate the equivalent impedance across two nodes of a network. We have shown how to use either a voltage or current source excitation and have given a probabilistic interpretation of both approaches. The remaining thing to do is to show that the probabilistic interpretations are consistent with each other.

When we connected a voltage source across nodes 0 and $n+1$ of a network we arrived at equation 2.2 for the equivalent admittance across the nodes. That equation is repeated here

$$Y_0 = \frac{I}{V_0} = \sum_{i=1}^{m}(1 - g_i)Y_{0i} \tag{8.1}$$

The sum is over all nodes connected to the source node (0) through admittances Y_{0i}. In terms of voltages, $g_i = V_i/V_0$ is a voltage transfer function. In probabilistic terms, g_i is the probability that a random walk starting at node i ends at node 0. Recall that the graph on which g_i is calculated only has transitions to node 0. There are no transitions away from 0. g_i is therefore the probability that when the walk ends at 0 it has not previously visited 0.

Connecting a current source across the same nodes, we arrived at equation 7.12 for the equivalent impedance across the nodes. That equation is repeated here

$$Z_0 = \frac{V_0}{I} = \frac{B_{00}}{\sum_{i=1}^{m} Y_{0i}} \tag{8.2}$$

The sum is over all nodes connected to the source node (0) through admittances Y_{0i} and, in probabilistic terms, B_{00} is the sum of the probabilities of random walks of all lengths that start and end at node 0.

To prove the consistency of these two equations we will equate them and show that it leads to a consistent probabilistic interpretation. Equating the two equations, we have

$$\sum_{i=1}^{m}(1 - g_i)Y_{0i} = \frac{\sum_{i=1}^{m} Y_{0i}}{B_{00}} \tag{8.3}$$

With a bit of rearranging we get

$$B_{00} = \frac{\sum_{i=1}^{m} Y_{0i}}{\sum_{i=1}^{m} Y_{0i} - \sum_{i=1}^{m} g_i Y_{0i}} \tag{8.4}$$

Now divide the top and bottom of this equation by $\sum_{i=1}^{m} Y_{0i}$ and write the result as follows

$$B_{00} = \frac{1}{1 - \sum_{i=1}^{m} p_{0i} g_i} \tag{8.5}$$

where p_{0i} is the probability of a transition from node 0 to i, given by

$$p_{0i} = \frac{Y_{0i}}{\sum_{j=1}^{m} Y_{0j}} \tag{8.6}$$

With g_i being the probability that a walk starts at node i and ends at 0 without any previous visits to 0, it is clear that the sum in the denominator of equation 8.5, $\sum_{i=1}^{m} p_{0i} g_i$, is the probability that a walk starting at node

0 returns there for the first time at the end of the walk. To see this, note that any such walk must begin by transitioning from 0 to one of the $i = 1, 2, \ldots, m$ nodes connected to it with probability p_{0i} and from there it will finally return to 0 with probability g_i.

Let's call this first return probability f_0, then we can write equation 8.5 as follows

$$B_{00} = \frac{1}{1 - f_0} \tag{8.7}$$

Does this equation make sense from a probabilistic standpoint? If we expand the equation in a power series then it becomes clear that it is probabilistically correct. Expanding in a power series we have

$$B_{00} = 1 + f_0 + f_0^2 + f_0^3 + \cdots \tag{8.8}$$

The 1 in the expansion represents a walk that ends at 0 without ever leaving. The f_0 term represents a walk that returns to 0 for the first time at the end. The f_0^2 term represents a walk that returns to 0 for the second time at the end and so on. These are the probabilities for all the ways that a walk can return to 0 and the sum of these probabilities should be equal to B_{00} as we have defined it.

Figure 8.1: Oliver Heaviside (1850-1925) invented the operational calculus (Laplace transform) upon which much of modern circuit analysis is based. Public domain image courtesy of Wikimedia Commons at https://commons.wikimedia.org/wiki/File:Oheaviside.jpg

9

RESISTIVE BRIDGE CIRCUIT WITH CURRENT SOURCE

For a current source example we will redo the bridge circuit problem with a current source in place of the voltage source as shown in figure 9.1. The problem is to find the voltages at nodes 0, 1, and 2.

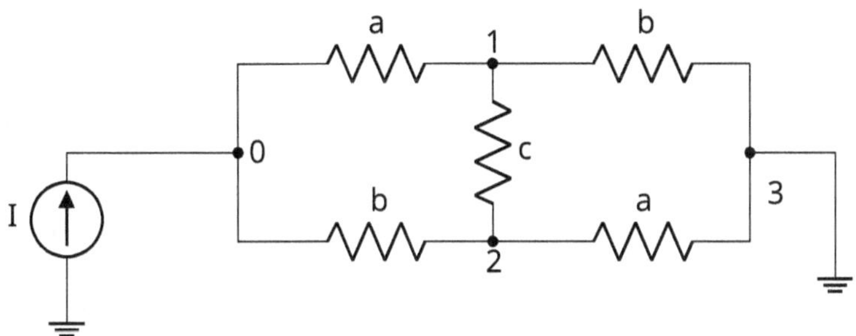

Figure 9.1: Resistive bridge circuit with current source.

The graph for the network is shown in figure 9.2. Note that we now have transitions from 0 to 1 and 2 whereas before we only had transitions from 1 and 2 to 0 as seen in figure 4.1.

The A matrix for this network is

$$A = \begin{bmatrix} 0 & p_{01} & p_{02} \\ p_{10} & 0 & p \\ p_{20} & p & 0 \end{bmatrix} \tag{9.1}$$

The values for p, p_{10} and p_{20} are given by equations 3.7, 3.8 and 3.9. The

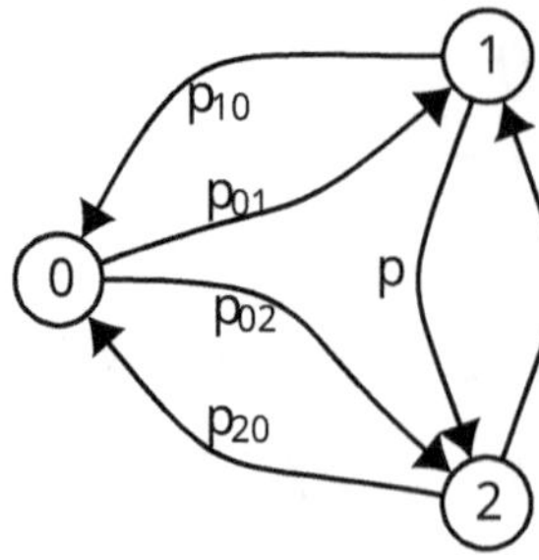

Figure 9.2: Bridge circuit graph.

values for p_{01} and p_{02} are given by

$$p_{01} = \frac{1/a}{1/a + 1/b} \tag{9.2}$$

$$p_{02} = \frac{1/b}{1/a + 1/b} \tag{9.3}$$

We will now solve the problem using $a = 4$, $b = 9$, $c = \sqrt{ab} = 6$ for which the probabilities are: $p = 6/19$, $p_{10} = 9/19$, $p_{20} = 4/19$, $p_{01} = 9/13$ and $p_{02} = 4/13$. The matrix form of the equation for the node voltages, $(I - A)\vec{V} = \vec{b}$, is then

$$\begin{bmatrix} 1 & -9/13 & -4/13 \\ -9/19 & 1 & -6/19 \\ -4/19 & -6/19 & 1 \end{bmatrix} \begin{bmatrix} V_0 \\ V_1 \\ V_2 \end{bmatrix} = \begin{bmatrix} 36I/13 \\ 0 \\ 0 \end{bmatrix} \tag{9.4}$$

and the solution, $\vec{V} = B\vec{b}$, where $B = (I - A)^{-1}$ is

$$\begin{bmatrix} V_0 \\ V_1 \\ V_2 \end{bmatrix} = \begin{bmatrix} 13/6 & 19/10 & 19/15 \\ 13/10 & 1463/650 & 361/325 \\ 13/15 & 361/325 & 1577/975 \end{bmatrix} \begin{bmatrix} 36I/13 \\ 0 \\ 0 \end{bmatrix} = \begin{bmatrix} 6I \\ 18I/5 \\ 12I/5 \end{bmatrix} \tag{9.5}$$

So the equivalent resistance between nodes 0 and 3 is $R_0 = V_0/I = 6$.

Now let's see how this can be solved using graphical techniques. If node 2 is eliminated in figure 9.2 and we use the probabilities listed above then we are left with the graph in figure 9.3. The $0 \to 0$ and $1 \to 1$ loop back probabilities are given by

$$p_{02}p_{20} = \frac{4}{13}\frac{4}{19} = \frac{16}{247} \tag{9.6}$$
$$p^2 = \frac{6}{19}\frac{6}{19} = \frac{36}{361} \tag{9.7}$$

The $0 \to 1$ and $1 \to 0$ transition probabilities are given by

$$p_{01} + p_{02}p = \frac{9}{13} + \frac{4}{13}\frac{6}{19} = \frac{15}{19} \tag{9.8}$$
$$p_{10} + p_{20}p = \frac{9}{19} + \frac{4}{19}\frac{6}{19} = \frac{195}{361} \tag{9.9}$$

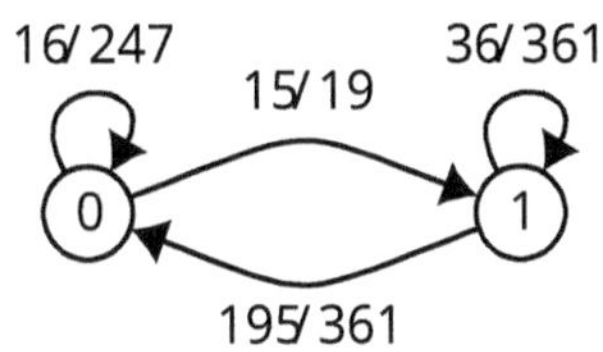

Figure 9.3: The graph in figure 9.2 with node 2 eliminated.

Finally we eliminate node 1 which leaves only node 0 as shown in figure 9.4. The loop back probability for node 0 is calculated as follows

$$\frac{16}{247} + \frac{\frac{15}{19}\frac{195}{361}}{1 - \frac{36}{361}} = \frac{7}{13} \tag{9.10}$$

From this we can calculate the sum of the probabilities for all walks that start and end at 0

$$B_{00} = \frac{1}{1 - \frac{7}{13}} = \frac{13}{6} \tag{9.11}$$

Figure 9.4: The graph in figure 9.3 with node 1 eliminated.

With B_{00} we can now calculate the voltage at 0

$$V_0 = \frac{B_{00} 36I}{13} = 6I \tag{9.12}$$

which is the same value we calculated above using matrix inversion.

To get V_1 we need B_{10} which is the sum of the probabilities of all walks that start at 1 and end at 0. That is equal to the probability of looping any number of times at 1 and then moving from 1 to 0 in figure 9.3 followed by the probability of finally ending at 0. So we have

$$B_{10} = \frac{195/361}{1 - \frac{36}{361}} B_{00} = \frac{13}{10} \tag{9.13}$$

and the voltage at 1 is then

$$V_1 = \frac{B_{10} 36I}{13} = \frac{18I}{5} \tag{9.14}$$

which is the same value we calculated above using matrix inversion.

To get V_2 we need to calculate B_{20} which we do by eliminating node 1 in figure 9.2. The resulting graph is shown in figure 9.5.

B_{20} is equal to the probability of moving from 2 to 0 in figure 9.5 followed by the probability of ending at 0. So we have

$$B_{20} = \frac{130/361}{1 - \frac{36}{361}} B_{00} = \frac{13}{15} \tag{9.15}$$

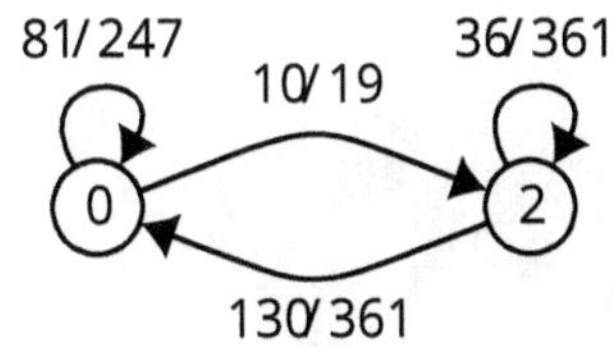

Figure 9.5: The graph in figure 9.2 with node 1 eliminated.

and the voltage at 2 is then

$$V_2 = \frac{B_{20} 36I}{13} = \frac{12I}{5} \tag{9.16}$$

which is the same value we calculated above using matrix inversion.

Figure 9.6: Léon Charles Thévenin (1857-1926) showed that a complex network composed of passive elements can be reduced to a voltage source in series with an impedance. Public domain image courtesy of Wikimedia Commons at https://commons.wikimedia.org/wiki/File:Léon_Charles_Thévenin.jpg

Figure 9.7: Edward Lawry Norton (1898-1983) showed that a complex network composed of passive elements can be reduced to a current source in parallel with an impedance. Public domain image courtesy of Wikimedia Commons at https://commons.wikimedia.org/wiki/File:Edward_Lawry_Norton.jpg

10

RESISTOR CUBE

If you connect resistors along the edges of a cube you get the network shown in figure 10.1. There are a total of 8 nodes connected by 12 resistors of equal value which we will set equal to 1. Measuring resistances across pairs of nodes, we find that there are only 3 unique values. We have R_{02}, the equivalent resistance across an edge of the cube, R_{07}, the equivalent resistance across the body diagonal of the cube and R_{05}, the equivalent resistance across the face diagonal of the cube.

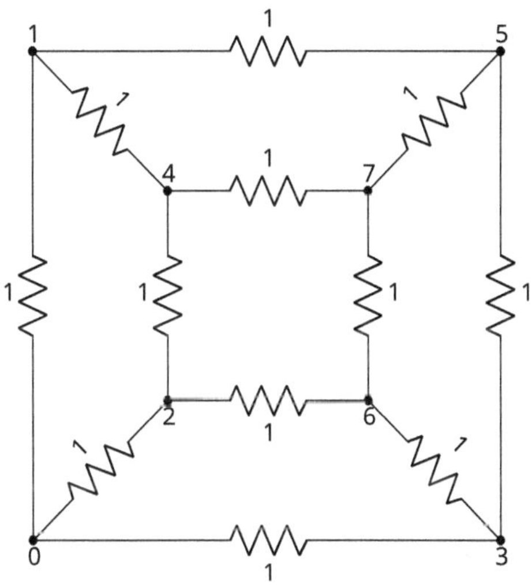

Figure 10.1: Resistor cube network.

To measure these resistances we will connect a current source to node 0 and ground node 2 for R_{02}, node 7 for R_{07} and node 5 for R_{05}. In all three cases it is clear from the symmetry of the network that we will have equal voltages at nodes 4 and 6, $V_4 = V_6$, and at nodes 1 and 3, $V_1 = V_3$. This means we can connect node 4 with 6 and node 1 with 3 without affecting the overall behavior of the network. If we connect nodes 4 and 6 we get the equivalent network shown in figure 10.2. If we now connect nodes 1 and 3 we get the equivalent network shown in figure 10.3.

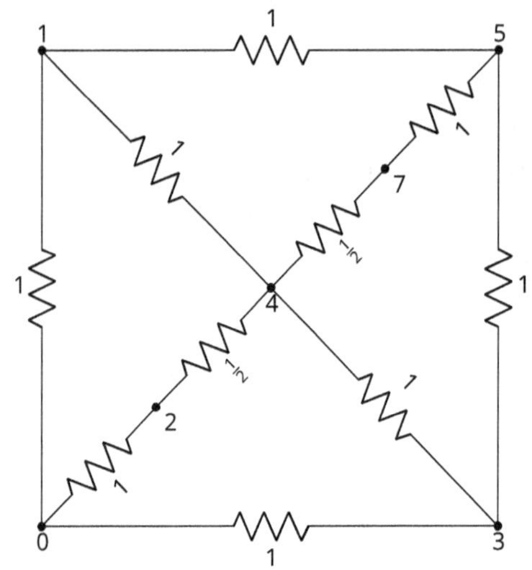

Figure 10.2: Resistor cube network with nodes 4 and 6 combined.

From this equivalent network R_{02} can be found using the standard method of combining series and parallel resistances. Following the node path $4-7-5-1$ there are 3 resistances in series that can be combined into a single resistor of value 2. This is in parallel with the resistance $1/2$ between nodes 4 and 1. Combining these parallel resistors gives a resistance of $2/5$. This leaves us with the network shown in figure 10.4 from which it is easy to calculate R_{02}. We have a resistance of $7/5$ in parallel with a resistance of 1 so we get $R_{02} = 7/12$.

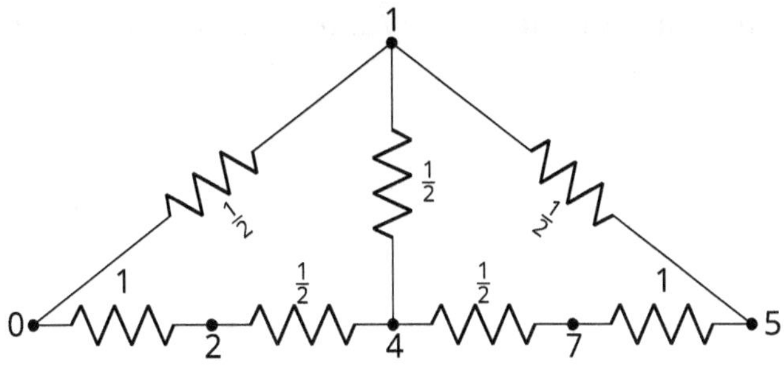

Figure 10.3: The network in figure 10.2 with nodes 1 and 3 combined.

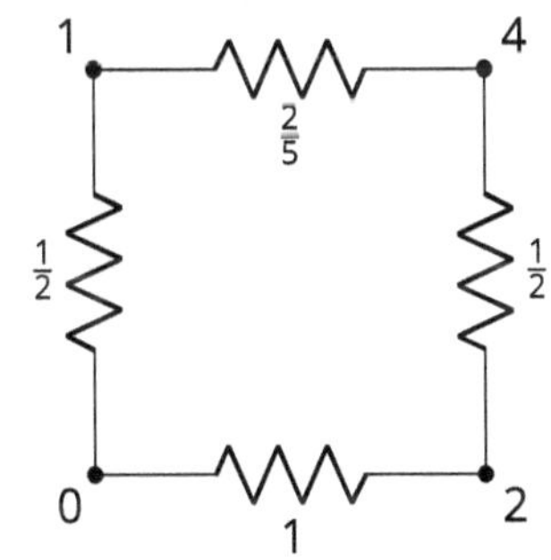

Figure 10.4: The network in figure 10.2 with nodes 1 and 3 combined.

To get R_{07} we eliminate nodes 5 and 2 in figure 10.3 by combining the resistors connected in series through those nodes to get the equivalent network shown in figure 10.5. This network is identical to the resistance bridge network shown in figure 9.1 with $a = 1/2$, $b = 3/2$ and $c = 1/2$. R_{07} can then be calculated using equation 3.20.

$$R_{07} = \frac{(a+b)c + 2ab}{a+b+2c} = \frac{5}{6} \tag{10.1}$$

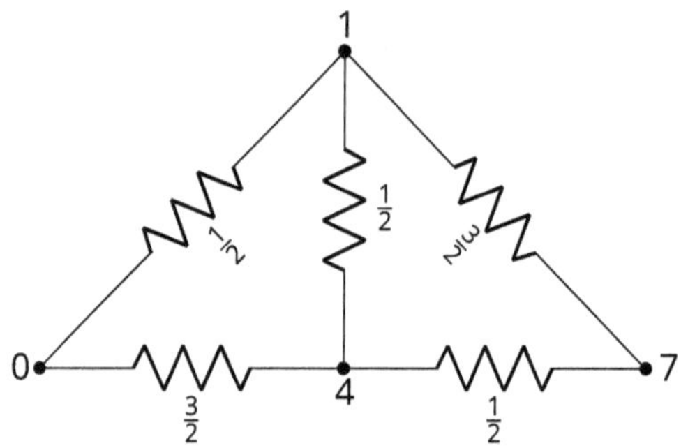

Figure 10.5: The network in figure 10.3 with nodes 2 and 5 eliminated.

To get R_{05} we eliminate nodes 2 and 7 in figure 10.3 by combining the resistors connected in series through those nodes to get the equivalent network shown in figure 10.6. This network is topologically identical to the resistance bridge network shown in figure 9.1 but it has different resistance values. The procedure for calculating the equivalent resistance is the same.

The graph for the network is shown in figure 10.7 and the A matrix for the network is

$$A = \begin{bmatrix} 0 & 3/4 & 1/4 \\ 1/3 & 0 & 1/3 \\ 1/5 & 3/5 & 0 \end{bmatrix} \tag{10.2}$$

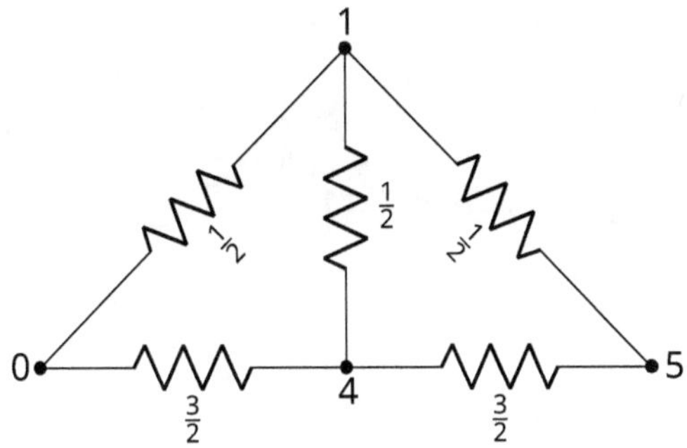

Figure 10.6: The network in figure 10.3 with nodes 2 and 7 eliminated.

The $B = (I - A)^{-1}$ matrix is

$$B = \begin{bmatrix} 2 & 9/4 & 5/4 \\ 1 & 19/8 & 25/24 \\ 1 & 15/8 & 15/8 \end{bmatrix} \tag{10.3}$$

and we get the equivalent resistance using equation 7.12

$$R_{05} = \frac{B_{00}}{\sum_j Y_{0j}} = \frac{2}{2 + 2/3} = \frac{3}{4} \tag{10.4}$$

This problem could also be solved using the node elimination techniques we used in previous problems. We leave it as an exercise to solve the problem by eliminating nodes 1 and 4 in the graph to get B_{00}.

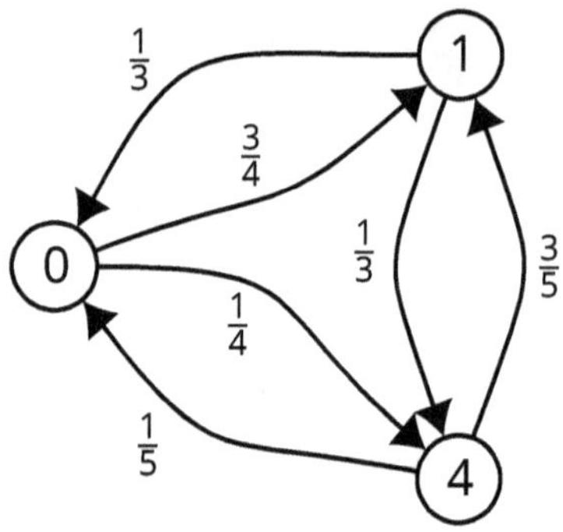

Figure 10.7: Graph for circuit in figure 10.6.

11

TWIN-T NOTCH FILTER

The circuit shown in figure 11.1 is called a twin-T notch filter. The filter input is V_0 at node 0 and the output is V_3 at node 3. We want to find the voltage transfer function $g_3 = V_3/V_0$ as a function of frequency and show that it goes to zero when the frequency is equal to $1/2\pi RC$.

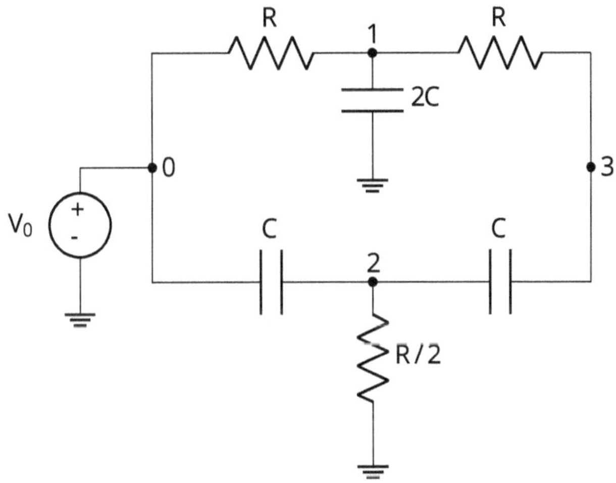

Figure 11.1: Twin-T notch filter network.

We will solve this problem using a purely graphical technique. The graph for the network is shown in figure 11.2. We are driving the network with a voltage source so the voltage at node 0 is fixed and there are only inward transitions, no outward transitions.

From the network diagram we get the following transition probabilities

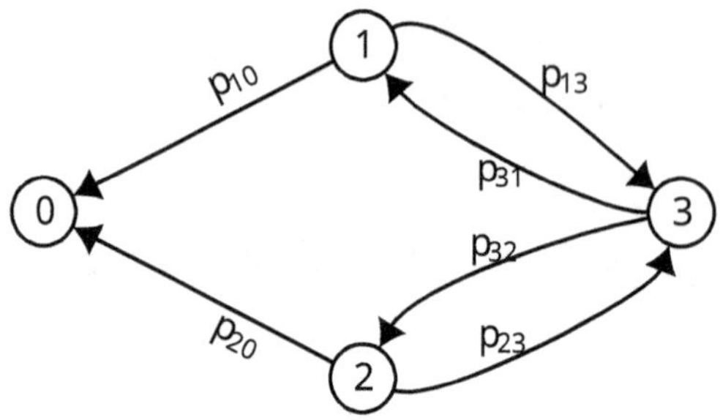

Figure 11.2: Twin-T notch filter graph.

coming out of node 1.

$$p_{10} = p_{13} = \frac{1/R}{2/R + 2Cs} = \frac{1}{2(RCs + 1)} \tag{11.1}$$

The transition probabilities coming out of node 2 are:

$$p_{20} = p_{23} = \frac{Cs}{2/R + 2Cs} = \frac{RCs}{2(RCs + 1)} \tag{11.2}$$

The transition probabilities coming out of node 3 are:

$$p_{31} = \frac{1/R}{1/R + Cs} = 2p_{10} \tag{11.3}$$

$$p_{32} = \frac{Cs}{1/R + Cs} = 2p_{20} \tag{11.4}$$

The transfer function $g_3 = V_3/V_0$ is the probability of a walk starting at node 3 and ending at node 0. We can find this probability by eliminating nodes 1 and 2 so that we are left with only nodes 0 and 3 as shown in figure

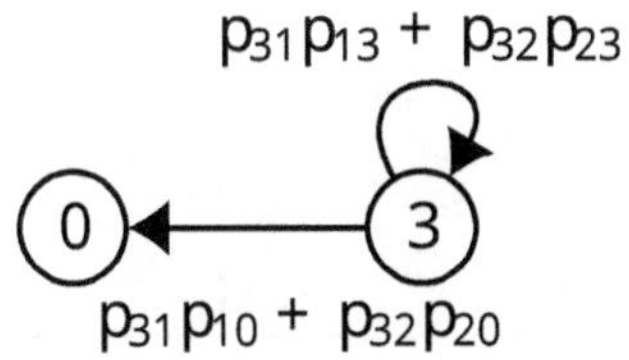

Figure 11.3: Reduced graph of figure 11.2.

11.3. From the figure we see that g_3 is equal to the loop back probability on node 3 followed by the transition to node 0.

$$g_3 = \frac{p_{31}p_{10} + p_{32}p_{20}}{1 - p_{31}p_{13} - p_{32}p_{23}} = \frac{2(p_{10}^2 + p_{20}^2)}{1 - 2(p_{10}^2 + p_{20}^2)} \tag{11.5}$$

Using the expressions for p_{10} and p_{20} given above we have

$$2(p_{10}^2 + p_{20}^2) = \frac{1}{2}\frac{(RCs)^2 + 1}{(RCs + 1)^2} \tag{11.6}$$

Substituting this into equation 11.5 and simplifying, we get

$$g_3 = \frac{(RCs)^2 + 1}{(RCs)^2 + 4RCs + 1} \tag{11.7}$$

To get the frequency response let $s = j\omega$, where $j = \sqrt{-1}$ then

$$g_3 = \frac{1 - (RC\omega)^2}{1 - (RC\omega)^2 + j4RC\omega} \tag{11.8}$$

Let $RC = 1/\omega_0$ then we can define the dimensionless variable $x = \omega/\omega_0$ and write the equation as follows

$$g_3(x) = \frac{1 - x^2}{1 - x^2 + j4x} \tag{11.9}$$

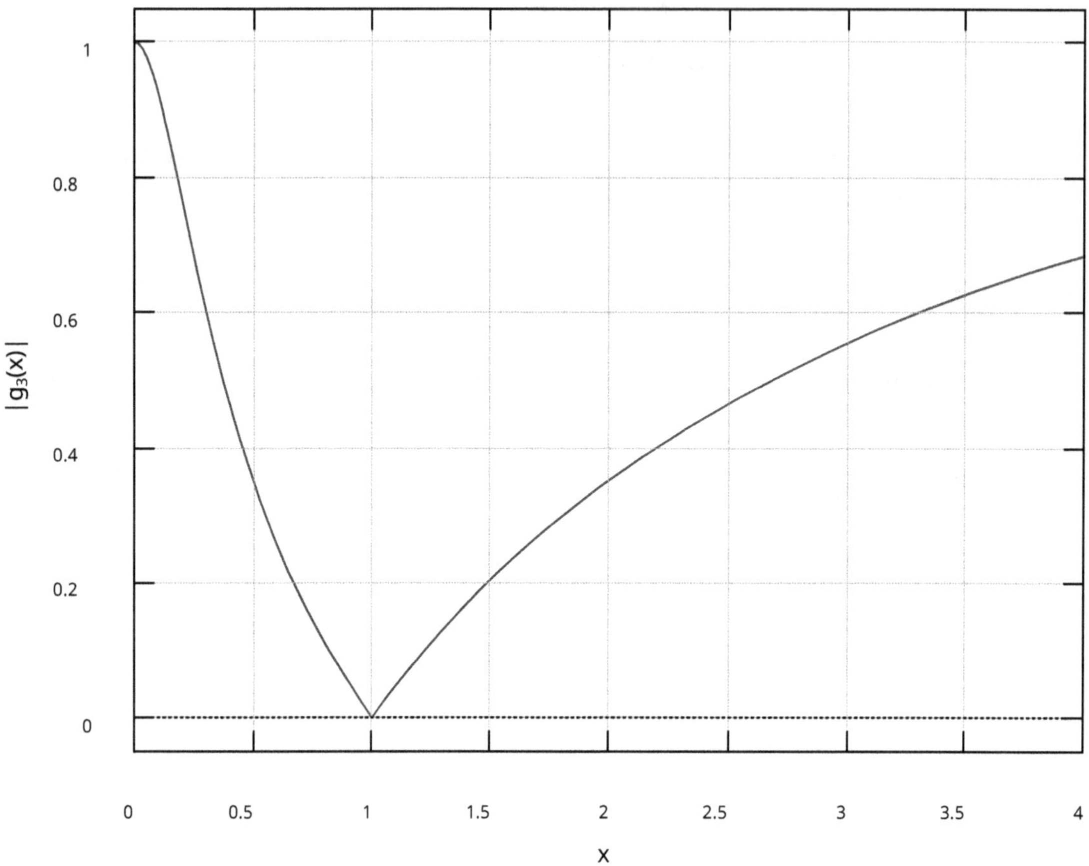

Figure 11.4: Twin-T frequency response.

The magnitude of the frequency response is then

$$|g_3(x)| = \frac{|1 - x^2|}{\sqrt{(1 - x^2)^2 + 16x^2}} \tag{11.10}$$

This clearly shows that the twin-T is a notch filter with a zero at $x = 1$ or $\omega = \omega_0 = 1/RC$. A plot of the magnitude of the frequency response is shown in figure 11.4.

One of the drawbacks of the twin-T notch filter is that it's not easy to tune. A notch filter that can be tuned using a single potentiometer is shown in figure 11.5 along with its graph. The input is at node 0 and the output is taken at node 1. Using the node elimination techniques we can easily find the transfer function from node 0 to node 1. In terms of the probabilities the transfer function is

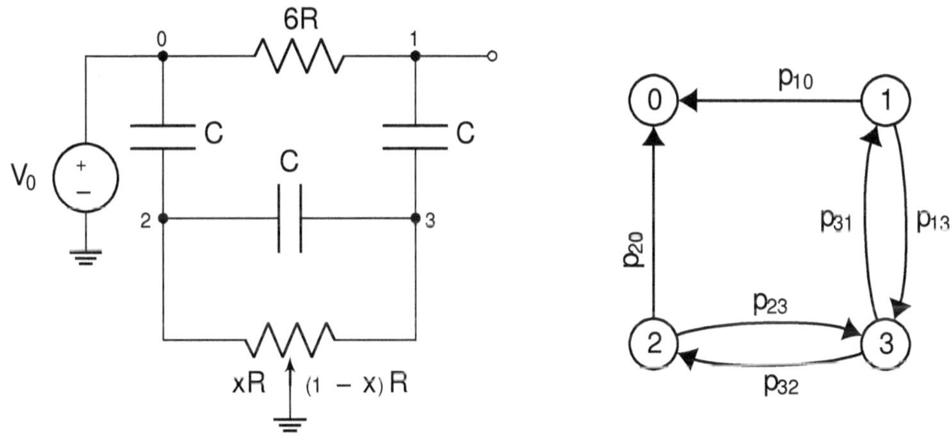

Figure 11.5: Tuneable notch filter.

$$g_1 = \frac{p_{10}(1 - p_{32}p_{23}) + p_{13}p_{32}p_{20}}{1 - p_{32}p_{23} - p_{13}p_{31}} \tag{11.11}$$

where the probabilities are given by

$$p_{10} = \frac{1}{6RCs + 1} \tag{11.12}$$

$$p_{13} = 1 - p_{10} \tag{11.13}$$

$$p_{20} = \frac{xRCs}{2xRCs + 1} \tag{11.14}$$

$$p_{23} = p_{20} \tag{11.15}$$

$$p_{31} = \frac{(1 - x)RCs}{2(1 - x)RCs + 1} \tag{11.16}$$

$$p_{32} = p_{31} \tag{11.17}$$

To find the notch frequency set $g_1 = 0$, let $s = j\omega$, and solve for ω. We find that the notch frequency is at

$$\omega = \frac{1}{RC\sqrt{3x(1 - x)}} \tag{11.18}$$

12

DIGITAL TO ANALOG CONVERSION

As a somewhat more general problem involving multiple voltage sources we will look at the four bit digital to analog conversion circuit shown in figure 12.1. The four voltage sources b_0, b_1, b_2, b_3, can have values that switch between 0 and 1. They are used to represent a four bit binary value. The output value of the circuit is taken from node 3. There should be a total of 16 possible output values corresponding to the 16 binary values represented by the b_i sources. In particular we want V_3 to be given by the following equation

$$V_3 = \frac{b_0 + 2b_1 + 4b_2 + 8b_3}{16} \tag{12.1}$$

This input-output relationship is shown in the following table.

$b_3b_2b_1b_0$	V_3	$b_0b_1b_2b_3$	V_3
0000	0	1000	8/16
0001	1/16	1001	9/16
0010	2/16	1010	10/16
0011	3/16	1011	11/16
0100	4/16	1100	12/16
0101	5/16	1101	13/16
0110	6/16	1110	14/16
0111	7/16	1111	15/16

To show that the circuit in figure 12.1 does satisfy this relationship we will start by using standard nodal analysis on nodes 0 thru 3. For node 0 we

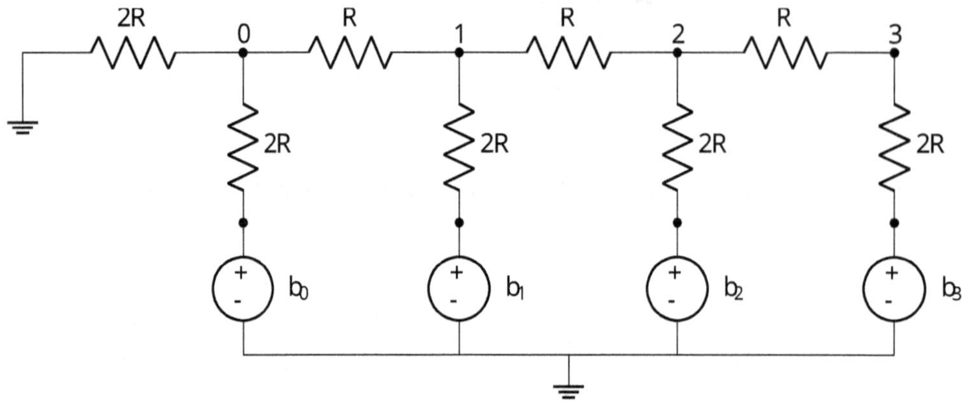

Figure 12.1: Four bit digital to analog converter.

have

$$\frac{V_0}{2R} + \frac{V_0 - b_0}{2R} + \frac{V_0 - V_1}{R} = 0 \tag{12.2}$$

which simplifies to

$$V_0 - \frac{1}{2}V_1 = \frac{1}{4}b_0 \tag{12.3}$$

For node 1 we have

$$\frac{V_1 - V_0}{R} + \frac{V_1 - b_1}{2R} + \frac{V_1 - V_2}{R} = 0 \tag{12.4}$$

which simplifies to

$$-\frac{2}{5}V_0 + V_1 - \frac{2}{5}V_2 = \frac{1}{5}b_1 \tag{12.5}$$

For node 2 we have

$$\frac{V_2 - V_1}{R} + \frac{V_2 - b_2}{2R} + \frac{V_2 - V_3}{R} = 0 \tag{12.6}$$

which simplifies to

$$-\frac{2}{5}V_1 + V_2 - \frac{2}{5}V_3 = \frac{1}{5}b_2 \tag{12.7}$$

For node 3 we have

$$\frac{V_3 - V_2}{R} + \frac{V_3 - b_3}{2R} = 0 \tag{12.8}$$

which simplifies to

$$-\frac{2}{3}V_2 + V_3 = \frac{1}{3}b_3 \tag{12.9}$$

Equations 12.3, 12.5, 12.7, 12.9 can be put in matrix form, $(I - A)\vec{V} = \vec{b}$, as follows

$$\begin{bmatrix} 1 & -1/2 & 0 & 0 \\ -2/5 & 1 & -2/5 & 0 \\ 0 & -2/5 & 1 & -2/5 \\ 0 & 0 & -2/3 & 1 \end{bmatrix} \begin{bmatrix} V_0 \\ V_1 \\ V_2 \\ V_3 \end{bmatrix} = \begin{bmatrix} b_0/4 \\ b_1/5 \\ b_2/5 \\ b_3/3 \end{bmatrix} \tag{12.10}$$

and the solution, $\vec{V} = B\vec{b}$, where $B = (I - A)^{-1}$, is

$$\begin{bmatrix} V_0 \\ V_1 \\ V_2 \\ V_3 \end{bmatrix} = \begin{bmatrix} 43/32 & 55/64 & 15/32 & 3/16 \\ 11/16 & 55/32 & 15/16 & 3/8 \\ 3/8 & 15/16 & 15/8 & 3/4 \\ 1/4 & 5/8 & 5/4 & 3/2 \end{bmatrix} \begin{bmatrix} b_0/4 \\ b_1/5 \\ b_2/5 \\ b_3/3 \end{bmatrix} \tag{12.11}$$

So $V_3 = b_0/16 + b_1/8 + b_2/4 + b_3/2$ which is the same as equation 12.1.

To solve the problem graphically, note that the A matrix in equation 12.10 represents the transition probability matrix for the linear graph shown in figure 12.2. To calculate V_3 we need the B_{30}, B_{31}, B_{32}, B_{33} elements of $B = (I - A)^{-1}$. These are the sums of the probabilities of all walks that start at 3 and end at 0, 1, 2 and 3 respectively.

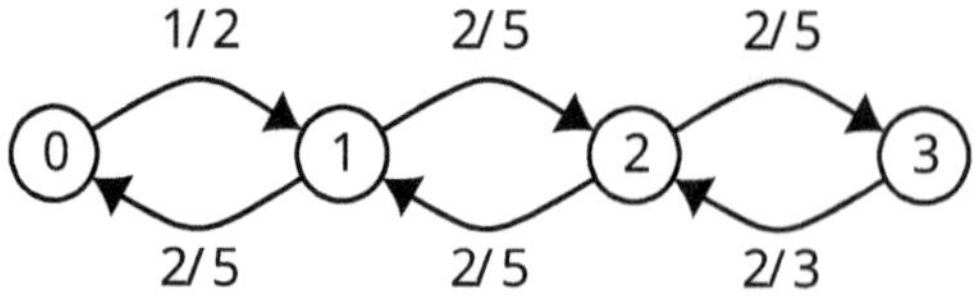

Figure 12.2: The graph for the circuit in figure 12.1.

To calculate B_{33} we have to eliminate, in order, nodes 0, 1, and 2, leaving only node 3. After eliminating node 0 we are left with the graph in figure 12.3. The loopback probability on node 1 comes from the transition from 1 to 0 and back again, $2/5 \cdot 1/2 = 1/5$.

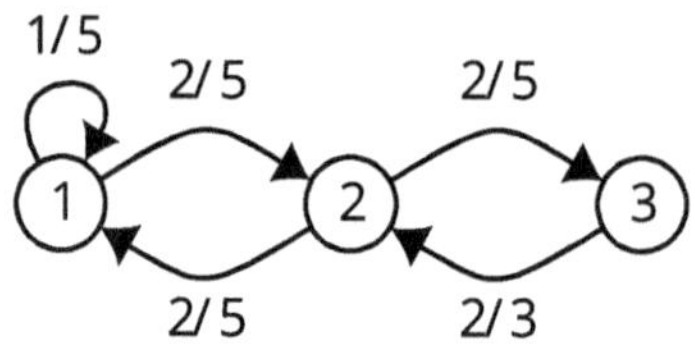

Figure 12.3: The graph in figure 12.2 with node 0 eliminated.

After eliminating node 1 we are left with the graph in figure 12.4. The loopback probability on node 2 comes from the transition from 2 to 1, the loop back on 1 and the transition back to 2 again. The calculation is

$$\frac{2/5 \cdot 2/5}{1 - 1/5} = 1/5 \tag{12.12}$$

After eliminating node 2 we are left with only node 3 as shown in figure 12.5. The loop back probability on 3 is calculated as follows

$$\frac{2/3 \cdot 2/5}{1 - 1/5} = 1/3 \tag{12.13}$$

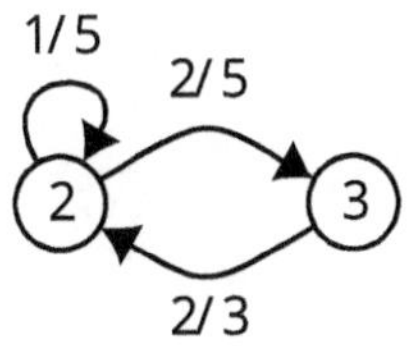

Figure 12.4: The graph in figure 12.3 with node 1 eliminated.

Figure 12.5: The graph in figure 12.4 with node 2 eliminated.

From this last graph we can write down the value of B_{33}

$$B_{33} = \frac{1}{1 - 1/3} = 3/2 \tag{12.14}$$

which is the same as the value in full B matrix in equation 12.11 that was found by inverting $(I - A)$.

B_{32} is the sum of the probabilities of all walks starting at 3 and ending at 2. Such walks can be decomposed into a walk starting and stopping at 3 followed by a transition to 2 and ending there. So we have

$$B_{32} = B_{33} \frac{2/3}{1 - 1/5} = 5/4 \tag{12.15}$$

Similarly for B_{31} and B_{30} we have

$$B_{31} = B_{32} \frac{2/5}{1 - 1/5} = 5/8 \tag{12.16}$$

$$B_{30} = B_{31} \frac{2}{5} = 1/4 \tag{12.17}$$

these all match the values in equation 12.11.

Study and in general the pursuit of truth and beauty is a sphere of activity in which we are permitted to remain children all our lives.

Albert Einstein

13

INFINITE NETWORKS

The random walk formalism can be extended to infinite networks such as the infinite length resistance ladder shown in figure 13.1. The series resistors have value R_1 and the shunt resistors have value R_2. We will find the equivalent resistance of the network using the voltage source method.

Using the same formalism employed in previous voltage source examples we can get equations for the nodes. For node 1 we get the equation $g_1 - p\,g_2 = p$. For node 2 we get the equation $-p\,g_1 + g_2 - p\,g_3 = 0$, and so on, where $g_i = V_i/V_0$ and the transition probability is

$$p = \frac{1/R_1}{2/R_1 + 1/R_2} = \frac{R_2}{R_1 + 2R_2} \tag{13.1}$$

All these equations together form an infinite tridiagonal matrix

$$\begin{bmatrix} 1 & -p & 0 & 0 & \cdots \\ -p & 1 & -p & 0 & \cdots \\ 0 & -p & 1 & -p & \cdots \\ \vdots & \vdots & \vdots & \vdots & \ddots \end{bmatrix} \begin{bmatrix} g_1 \\ g_2 \\ g_3 \\ \vdots \end{bmatrix} = \begin{bmatrix} p \\ 0 \\ 0 \\ \vdots \end{bmatrix} \tag{13.2}$$

There are techniques for solving such an equation but we will instead use the graphical technique. The graph for the network is shown in figure 13.2.

There are an infinite number of nodes in the graph but we can still use node elimination to eliminate all but nodes 0 and 1. This will reduce the graph to that shown in figure 13.3.

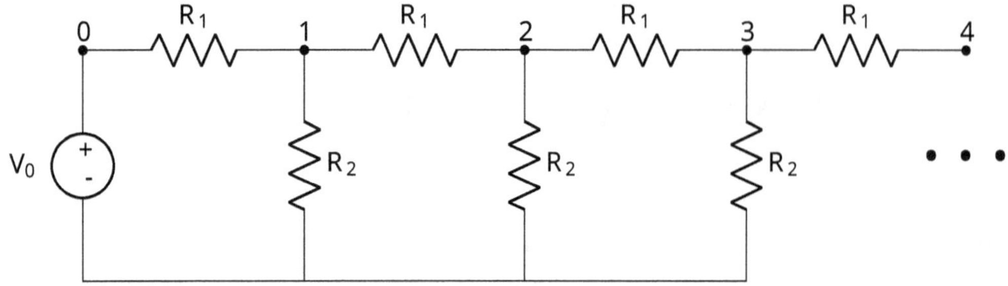

Figure 13.1: An infinite length resistance ladder.

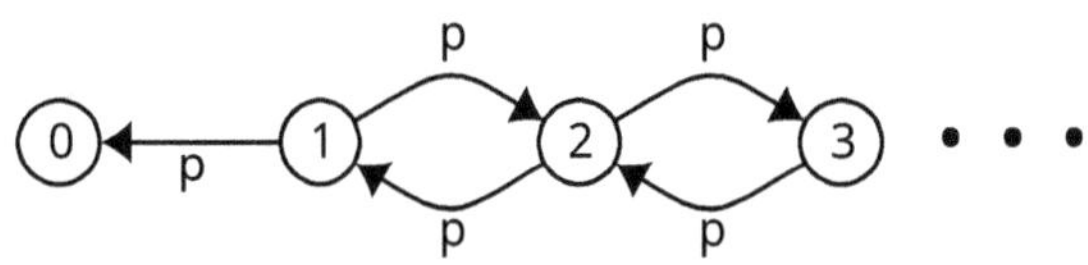

Figure 13.2: The graph for the infinite length resistance ladder of figure 13.1.

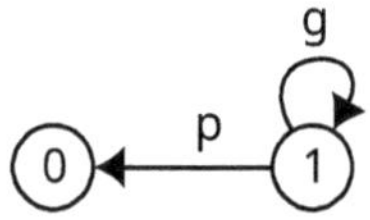

Figure 13.3: Reduced graph of figure 13.2.

The loop back probability, g, on node 1 is an infinite continued fraction of the following form

$$g = \cfrac{p^2}{1 - \cfrac{p^2}{1 - \cfrac{p^2}{1 - \cfrac{p^2}{1 - \cdots}}}} \tag{13.3}$$

To convince yourself that the probability has this form, try calculating the corresponding probability for a graph with only the nodes explicitly shown in figure 13.2.

Since this is an infinite continued fraction we can see that the second term in the denominator is itself equal to the g. We can therefor write the equation in the following form

$$g = \frac{p^2}{1 - g} \tag{13.4}$$

In this form we can solve for g. We find that g is given by

$$g = \frac{1 - \sqrt{1 - 4p^2}}{2} \tag{13.5}$$

Now from the reduced graph in figure 13.3 we can write down the expression for g_1

$$g_1 = \frac{V_1}{V_0} = \frac{p}{1 - g} \tag{13.6}$$

Comparing this with equation 13.4 we see that $g_1 = g/p$ or

$$g_1 = \frac{1 - \sqrt{1 - 4p^2}}{2p} \tag{13.7}$$

With g_1 we can now find the equivalent resistance of the network. If I is the current entering node 0 from the voltage source then

$$I = \frac{V_0 - V_1}{R_1} \tag{13.8}$$

Dividing both sides of this equation by V_0 gives us an expression for the equivalent conductance of the network in terms of R_1 and $g_1 = V_1/V_0$. Invert the equation and we have the equivalent resistance of the network.

$$R_0 = \frac{V_0}{I} = \frac{R_1}{1 - g_1} \tag{13.9}$$

Substituting the expression for g_1 and simplifying we get

$$R_0 = R_1 \left(\frac{1 + \sqrt{1 + 4\alpha}}{2} \right) \tag{13.10}$$

where $\alpha = R_2/R_1$.

When $R_1 = R_2 = R$ then $\alpha = 1$ and the equivalent resistance is proportional to the golden ratio.

$$R_0 = R \left(\frac{1 + \sqrt{5}}{2} \right) \approx 1.6180339887 \cdot R \tag{13.11}$$

Another infinite network is shown in figure 13.4. The network has an infinite number of rungs to the right and left of the current source that is driving the network at node 0. We want to find the equivalent resistance looking into node 0.

The corresponding graph for this network is shown in figure 13.5. This is once again an infinite graph with nodes going to plus infinity to the

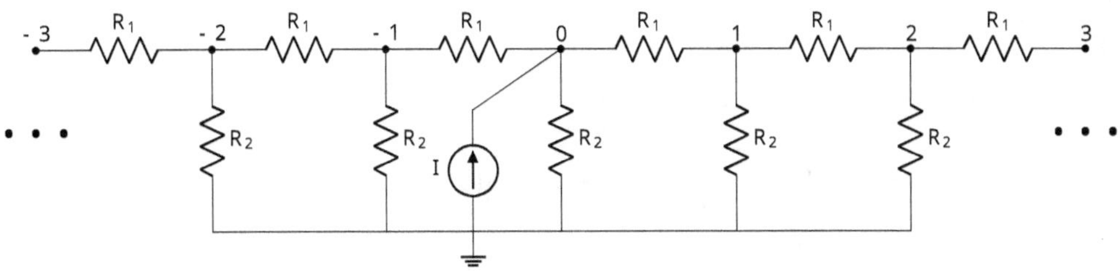

Figure 13.4: Two way infinite resistance ladder with current source.

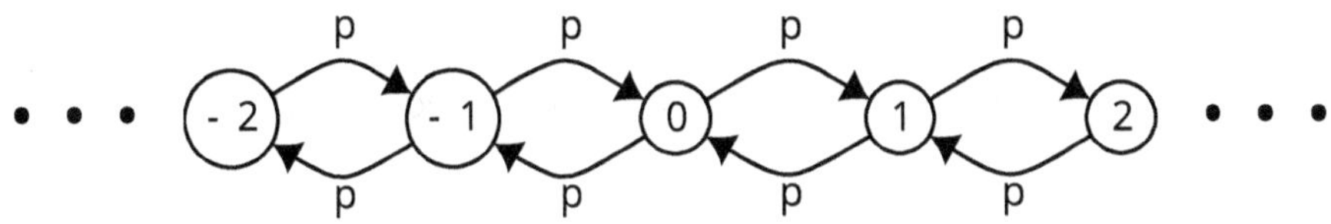

Figure 13.5: Graph of the network in figure 13.4.

right and negative infinity to the left. All transitions in the graph have probability p given by

$$p = \frac{1/R_1}{2/R_2 + 1/R_2} = \frac{R_2}{R_1 + 2R_2} \tag{13.12}$$

The adjacency matrix for the graph is an infinite tridiagonal matrix. Luckily we do not have to calculate the entire $B = (I - A)^{-1}$ matrix. All we need is the B_{00} element of the matrix which is the sum of the probabilities for all walks that start and end at node 0. We can find that probability using graphical methods.

Suppose we eliminated all the nodes to the right of node 1 and all the nodes to the left of node -1 then we would be left with the graph in figure 13.6. The remaining nodes 1 and -1 have loop back probabilities that represent traveling among the eliminated nodes and returning. Clearly, the loop back probability g in figure 13.5 must be the same as the loop back probability g in figure 13.3 that was calculated in the previous problem and is given by eq. 13.5.

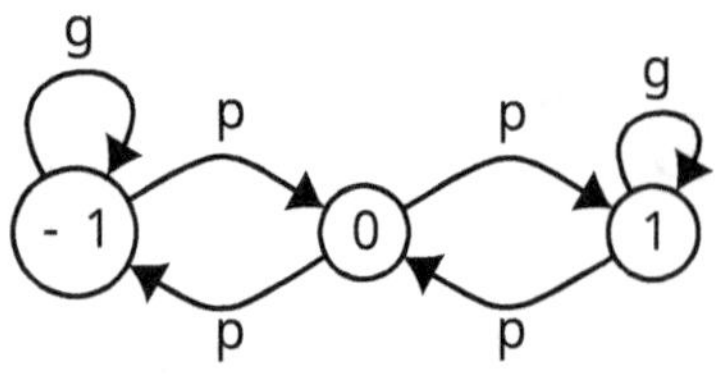

Figure 13.6: Reduced graph of figure 13.5.

Finally we can eliminate nodes -1 and 1, leaving only node 0 with a loop back probability equal to

$$p_{00} = \frac{2p^2}{1-g} \tag{13.13}$$

From this we can write down the expression for B_{00}.

$$B_{00} = \frac{1}{1-p_{00}} = \frac{1}{\sqrt{1-4p^2}} \tag{13.14}$$

The voltage induced at node 0 by the current source is then

$$V_0 = \frac{B_{00}I}{2/R_1 + 1/R_2} \tag{13.15}$$

Dividing both sides of this equation by I we get the equivalent resistance looking into node 0

$$R_0 = \frac{V_0}{I} = \frac{B_{00}}{2/R_1 + 1/R_2} \tag{13.16}$$

With some substitutions and simplification we have

$$\frac{R_0}{R_1} = \frac{p}{\sqrt{1-4p^2}} = \frac{\alpha}{\sqrt{1+4\alpha}} \tag{13.17}$$

where $\alpha = R_2/R_1$.

As an example, for $R_1 = R_2 = R$ we have $p = 1/3$ or $\alpha = 1$ and

$$\frac{R_0}{R} = \frac{1}{\sqrt{5}} \tag{13.18}$$

As an exercise show that this is equal to half the resistance in equation 13.11 in parallel with resistance R. Do you see why this is true?

The mind is not a vessel to be filled, but a fire to be kindled.

Plutarch

14

TRANSMISSION LINE

When the wavelength of signals in a circuit is much larger than the size of the circuit, the connections between circuit components have little effect. As wavelengths get smaller and become comparable to the size of the circuit the effect of the connections can no longer be ignored and they have to be modeled as transmission lines. A transmission line is essentially a waveguiding structure. Some common examples are coaxial cables and the twin lead used in TV and antenna connections. Traces on circuit boards can also be modeled as transmission lines. What all these structures have in common is that there is a distributed inductance and capacitance down the length of the line. The data sheet for a particular coaxial cable will often list the capacitance per length of the cable and a parameter called the impedance from which the inductance per length can be calculated. If c and l are the capacitance and inductance per unit length then an infinitesimal length dx of the line will have capacitance $C = c\,dx$ and inductance $L = l\,dx$ and the entire transmission line can be modeled using the infinite circuit shown in figure 14.1.

This circuit is topologically identical to the infinite length resistance ladder of figure 13.1 and its graph is identical to the graph shown in figure 13.2. Since the graph is identical the solution for $g_1 = V_1/V_0$ will also be given by equation 13.7 which we repeat below.

$$g_1 = \frac{1 - \sqrt{1 - 4p^2}}{2p} \tag{14.1}$$

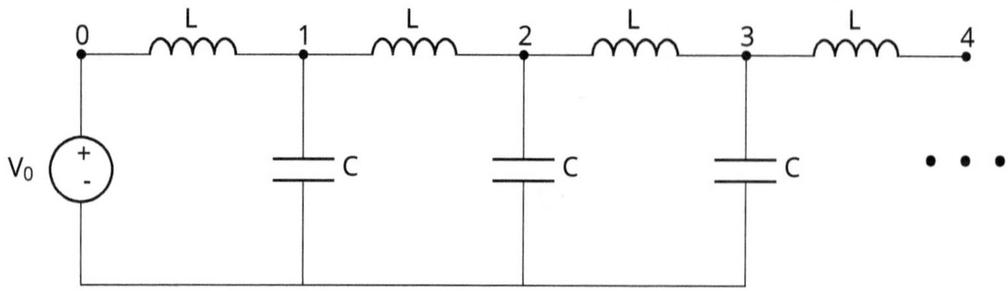

Figure 14.1: Transmission line model.

But in this case the probability p is equal to

$$p = \frac{1}{2 - \omega^2 LC} \tag{14.2}$$

or if we let $LC = 1/\omega_0^2$ and $x = \omega/\omega_0$ then we can write p as

$$p = \frac{1}{2 - x^2} \tag{14.3}$$

We can use g_1 to get the equivalent impedance looking into node 0 of the network. The current entering node 0 is $I = (V_0 - V_1)/j\omega L$ so the impedance is

$$Z_0 = \frac{V_0}{I} = \frac{V_0 j\omega L}{V_0 - V_1} = \frac{j\omega L}{1 - g_1} \tag{14.4}$$

with g_1 given by equation 14.1 and p given by equation 14.2 we get

$$Z_0 = j\omega\frac{L}{2}\left(1 - \sqrt{1 - \frac{4}{x^2}}\right) \tag{14.5}$$

If we split the inductance between nodes 0 and 1 in half then leaving out the first half, the impedance of the rest of the circuit is

$$Z_0' = Z_0 - j\omega\frac{L}{2} = -j\omega\frac{L}{2}\sqrt{1 - \frac{4}{x^2}} \tag{14.6}$$

For $x < 2$ or $\omega < 2\omega_0$ the impedance is real so it absorbs energy, while for $x > 2$ or $\omega > 2\omega_0$ the impedance is a pure negative imaginary number so the entire transmission line looks like a capacitance that simply stores energy. Overall then the transmission line looks like a low pass filter. When $\omega < 2/\sqrt{LC}$ energy is propagated down the line away from the source. When $\omega > 2/\sqrt{LC}$ energy is stored in the line and not propagated.

We can see this more clearly by looking at the ratio of the voltage from one node to the next. In general the voltage at node $n+1$ is related to the voltage at node n as follows: $V_{n+1} = V_n - j\omega L I_n$. Since this is an infinite network the impedance looking into each node will be Z_0 therefor $I_n = V_n/Z_0$. Substituting this into the previous equation we see that the voltage ratio from one node to the next remains constant.

$$\alpha = \frac{V_{n+1}}{V_n} = 1 - \frac{j\omega L}{Z_0} \tag{14.7}$$

With Z_0 given by equation 14.5 and with some simplification we get

$$\alpha = -\frac{1+\sqrt{1-4/x^2}}{1-\sqrt{1-4/x^2}} \tag{14.8}$$

For $4/x^2 < 1$, $x > 2$, $\omega > 2/\sqrt{LC}$, α will be real with magnitude less than 1, so the voltage decays going down the line. For $4/x^2 > 1$, $x < 2$, $\omega < 2/\sqrt{LC}$, α will be complex with magnitude equal to 1 and phase $\phi = 2\tan^{-1}\sqrt{1-4/x^2}$, so $\alpha = -e^{j\phi}$ and the voltage propagates unattenuated down the line with only a change in phase.

The propagation constant in equation 14.8 can also be found from the network graph. We have $\alpha = V_{n+1}/V_n = g_{n+1}/g_n$ and from the graph we can see that the g_n must obey the following recurrence equation.

$$g_n = pg_{n-1} + pg_{n+1} \tag{14.9}$$

Linear recurrence equations like this can always be solved using the ansatz $g_n = z^n$. Substituting this into equation 14.9 we find that z must satisfy $pz^2 - z + p = 0$. The solutions to this equation are

$$z = \frac{1 \pm \sqrt{1-4p^2}}{2p} \tag{14.10}$$

Since g_1 is given by equation 14.1, we choose the negative solution to satisfy $g_1 = z$. Therefor we have $g_n = g_1^n$ and $\alpha = g_{n+1}/g_n = g_1$. With a bit of algebra we can verify that $\alpha = g_1$ is the same as equation 14.8.

15

DELTA - WYE TRANSFORMATION

Let's start with one of the simplest transformations in electrical network theory: combining two series connected resistors into a single resistor as shown in figure 15.1.

Figure 15.1: Combining two series connected resistors.

If the two series resistors have values S_1 and S_2 then the combined resistor will have value $R = S_1 + S_2$.

In the probabilistic formulation we have

$$p_{ba} = \frac{S_2}{S_1 + S_2} \tag{15.1}$$

$$p_{bc} = \frac{S_1}{S_1 + S_2} \tag{15.2}$$

Then we can write

$$\frac{1}{R} = \frac{1}{S_1 + S_2} = \frac{p_{bc}}{S_1} = \frac{p_{ba}}{S_2} \tag{15.3}$$

Now let's see how to apply this to the more complicated delta-wye transformation. This transformation is shown in figure 15.2.

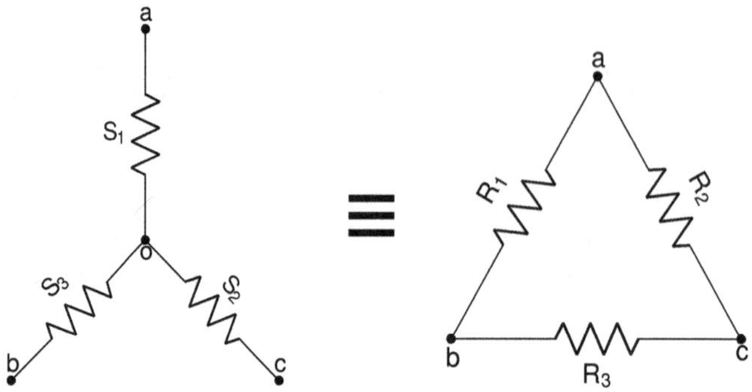

Figure 15.2: Delta-Wye transformation.

We want the resistor values to be such that the two configurations are indistinguishable. This means that the resistance measured between any two terminals must be the same for both circuits. Let's first look at this using standard network theory. If the resistance measured across terminals a and b is to be the same for both circuits then we must have

$$\frac{1}{S_1 + S_3} = \frac{1}{R_1} + \frac{1}{R_2 + R_3} \tag{15.4}$$

For the resistance across terminals a and c to be the same we must have

$$\frac{1}{S_1 + S_2} = \frac{1}{R_2} + \frac{1}{R_1 + R_3} \tag{15.5}$$

For the resistance across terminals b and c to be the same we must have

$$\frac{1}{S_2 + S_3} = \frac{1}{R_3} + \frac{1}{R_1 + R_2} \tag{15.6}$$

These last three equations can be put into matrix form as follows

$$\begin{bmatrix} 0 & 1 & 1 \\ 1 & 0 & 1 \\ 1 & 1 & 0 \end{bmatrix} \begin{bmatrix} S_1 \\ S_2 \\ S_3 \end{bmatrix} = \begin{bmatrix} R_3(R_1 + R_2) \\ R_1(R_2 + R_3) \\ R_2(R_1 + R_3) \end{bmatrix} \frac{1}{R_1 + R_2 + R_3} \tag{15.7}$$

Solving the equation for the S_i values we get

$$\begin{bmatrix} S_1 \\ S_2 \\ S_3 \end{bmatrix} = \frac{1}{2} \begin{bmatrix} -1 & 1 & 1 \\ 1 & -1 & 1 \\ 1 & 1 & -1 \end{bmatrix} \begin{bmatrix} R_3(R_1 + R_2) \\ R_1(R_2 + R_3) \\ R_2(R_1 + R_3) \end{bmatrix} \frac{1}{R_1 + R_2 + R_3} \tag{15.8}$$

Multiplying out the matrix we get

$$S_1 = \frac{R_1 R_2}{R_1 + R_2 + R_3} \tag{15.9}$$

$$S_2 = \frac{R_2 R_3}{R_1 + R_2 + R_3} \tag{15.10}$$

$$S_3 = \frac{R_1 R_3}{R_1 + R_2 + R_3} \tag{15.11}$$

Given R_1, R_2 and R_3, these equations can be used to find the values of S_1, S_2 and S_3, that make the two circuits equivalent. To go the other way we need equations for the R_i values in terms of the S_i values. These equations can be found by inverting the above equations. A simpler way is to use the probabilistic formulation for combining series resistances that we showed at the beginning of this chapter.

Using the probabilistic formulation we have

$$\frac{1}{R_1} = \frac{p_{ob}}{S_1} = \frac{p_{oa}}{S_3} \tag{15.12}$$

$$\frac{1}{R_2} = \frac{p_{oc}}{S_1} = \frac{p_{oa}}{S_2} \tag{15.13}$$

$$\frac{1}{R_3} = \frac{p_{ob}}{S_2} = \frac{p_{oc}}{S_3} \tag{15.14}$$

where the probabilities are given by

$$p_{oa} = \frac{1/S_1}{1/S_1 + 1/S_2 + 1/S_3} = \frac{S_2 S_3}{S_1 S_2 + S_1 S_3 + S_2 S_3} \tag{15.15}$$

$$p_{ob} = \frac{1/S_3}{1/S_1 + 1/S_2 + 1/S_3} = \frac{S_1 S_2}{S_1 S_2 + S_1 S_3 + S_2 S_3} \tag{15.16}$$

$$p_{oc} = \frac{1/S_2}{1/S_1 + 1/S_2 + 1/S_3} = \frac{S_1 S_3}{S_1 S_2 + S_1 S_3 + S_2 S_3} \tag{15.17}$$

Substituting these probabilities into equations 15.12, 15.13, and 15.14 we get

$$R_1 = \frac{S_1 S_2 + S_1 S_3 + S_2 S_3}{S_2} \tag{15.18}$$

$$R_2 = \frac{S_1 S_2 + S_1 S_3 + S_2 S_3}{S_3} \tag{15.19}$$

$$R_3 = \frac{S_1 S_2 + S_1 S_3 + S_2 S_3}{S_1} \tag{15.20}$$

As an example, if we have a wye configuration with $S_1 = 1$, $S_2 = 2$, and $S_3 = 3$, then the R_i values for the delta configuration will be $R_1 = 11/2$, $R_2 = 11/3$, and $R_3 = 11$. This makes $R_{ab} = 4$, $R_{ac} = 3$, and $R_{bc} = 5$ for both the delta and wye configurations.

The delta-wye transformation is just one example of what are called star-mesh transformations. Another example involving four terminals is shown in figure 15.3.

Using the same procedure as we did with the delta-wye transformation, you should be able to show that the R_i resistances are given by

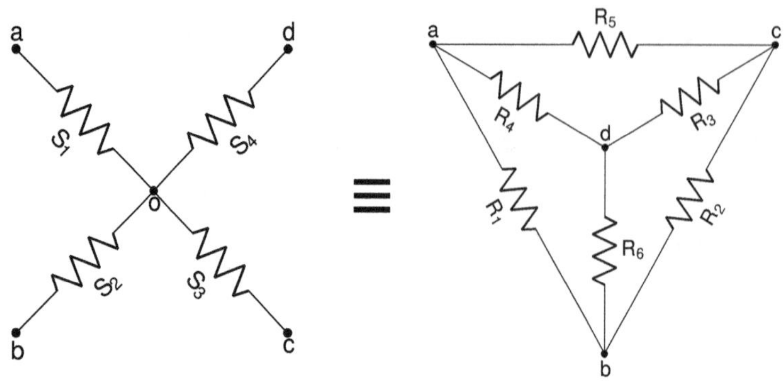

Figure 15.3: Four terminal Star-Mesh transformation.

$$R_1 = \frac{D}{S_3 S_4} \tag{15.21}$$

$$R_2 = \frac{D}{S_1 S_4} \tag{15.22}$$

$$R_3 = \frac{D}{S_1 S_2} \tag{15.23}$$

$$R_4 = \frac{D}{S_2 S_3} \tag{15.24}$$

$$R_5 = \frac{D}{S_2 S_4} \tag{15.25}$$

$$R_6 = \frac{D}{S_1 S_3} \tag{15.26}$$

$$D = S_1 S_2 S_3 + S_1 S_2 S_4 + S_1 S_3 S_4 + S_2 S_3 S_4 \tag{15.27}$$

Note that since there are more R_i resistances than S_i resistances, the transformation is in general not invertible. In other words, given the six R_i resistances for the mesh network, you cannot always find S_i resistances for the star network that will make the two networks equivalent.

16

OPERATIONAL AMPLIFIERS

Circuits involving operational amplifiers (op amps) can also be modeled using graphical techniques. The schematic symbol for an op amp and its corresponding graph representation are shown in figure 16.1.

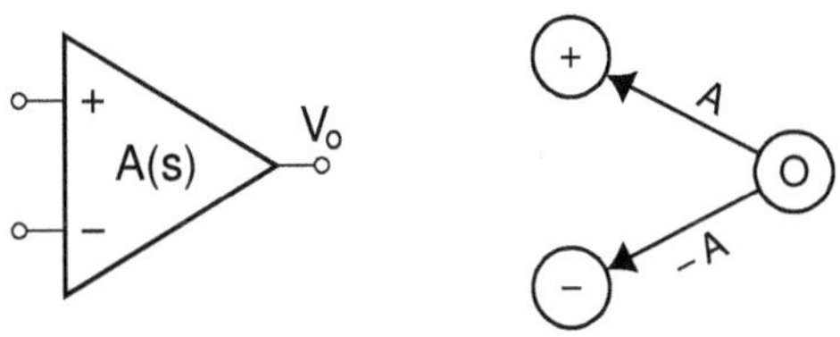

Figure 16.1: Op amp schematic symbol and its graph representation.

The transfer function is $V_0(s) = A(s)(V_+ - V_-)$. For most applications, the gain, $A(s)$, can be taken to be a one pole rolloff model

$$A(s) = \frac{A_0\, \omega_0}{s + \omega_0} \tag{16.1}$$

A_0 is the DC open loop gain and w_0 is the open loop 3 dB frequency. At frequencies much less than ω_0 the gain will be approximately A_0 which is typically on the order of 10^5. This means that when $\omega \ll \omega_0$ we can often take the limit $A(s) \to \infty$ to simplify the equations.

The three simplest op amp circuits are the inverting, noninverting and difference amplifiers. We will analyse them using the graphical model in this chapter and cover more complex op amp circuits in subsequent chapters.

Inverting Amplifier

The inverting amplifier and its graphical model are shown in figure 16.2.

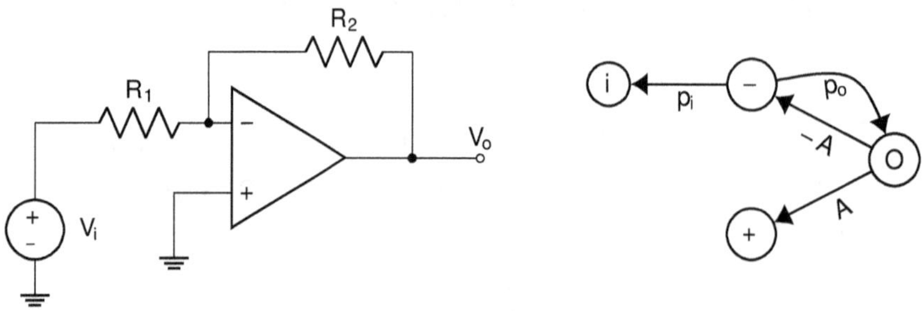

Figure 16.2: The inverting amplifier and its graphical model.

Since node + is at ground we can ignore it and just eliminate node −. This leaves the graph shown in figure 16.3.

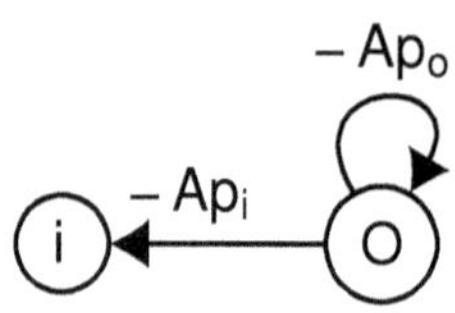

Figure 16.3: Simplified inverting amplifier graph.

From this we can just read off the transfer function

$$g_o = \frac{-Ap_i}{1 + Ap_o} = \frac{V_0}{V_i} \tag{16.2}$$

where

$$p_i = \frac{1/R_1}{1/R_1 + 1/R_2} = \frac{R_2}{R_1 + R_2} \tag{16.3}$$

$$p_o = \frac{1/R_2}{1/R_1 + 1/R_2} = \frac{R_1}{R_1 + R_2} \tag{16.4}$$

In the limit as $A \to \infty$ we have

$$g_o = -\frac{p_i}{p_o} = -\frac{R_2}{R_1} \tag{16.5}$$

Non-Inverting Amplifier

The non-inverting amplifier and its graph model are shown in figure 16.4.

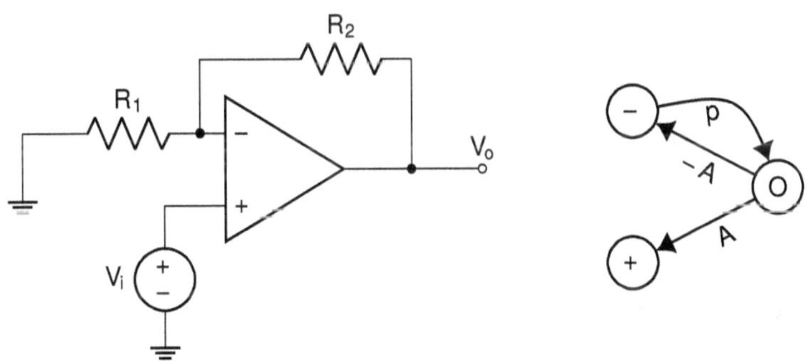

Figure 16.4: The non-inverting amplifier and its graphical model.

In this case we need to eliminate node "-" so we are left with the graph shown in figure 16.5.

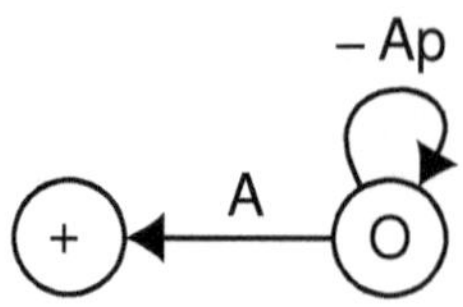

Figure 16.5: Simplified non-inverting amplifier graph.

From the graph we can read off the transfer function. We have

$$g_0 = \frac{V_o}{V_i} = \frac{A}{1 + Ap} \tag{16.6}$$

where

$$p = \frac{1/R_2}{1/R_1 + 1/R_2} = \frac{R_1}{R_1 + R_2} \tag{16.7}$$

In the limit as $A \to \infty$ we have

$$g_0 = \frac{1}{p} = \frac{R_1 + R_2}{R_1} = 1 + \frac{R_2}{R_1} \tag{16.8}$$

Difference Amplifier

The difference amplifier and its graph model are shown in figure 16.6.

In this case we need to eliminate nodes "+" and "-". We are left with the graph shown in figure 16.7.

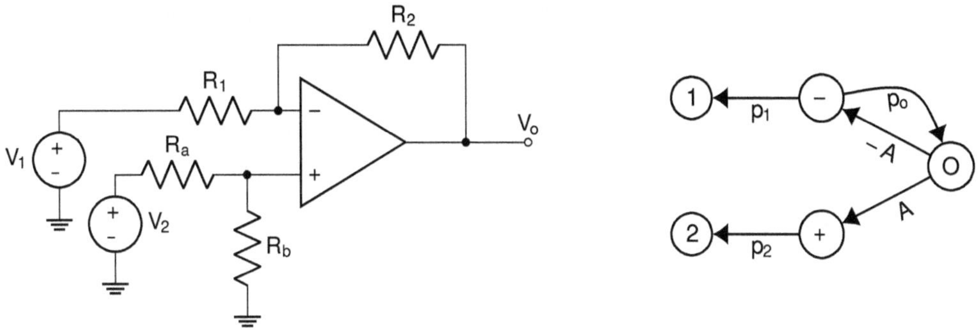

Figure 16.6: The difference amplifier and its graphical model.

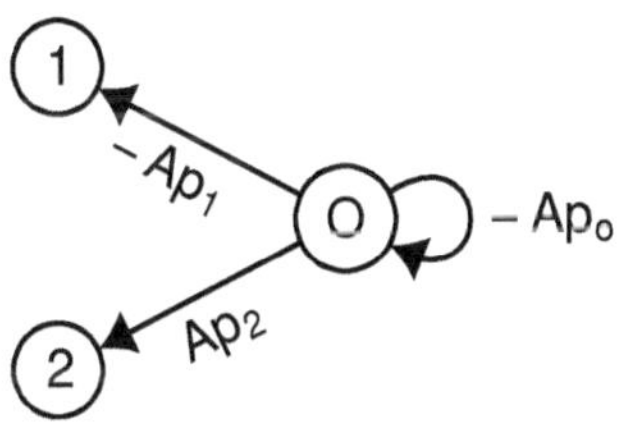

Figure 16.7: Simplified difference amplifier graph.

V_o depends on V_1 and V_2 so we use the principle of superposition and look at the transfer function due to nodes 1 and 2 separately. For node 1 we have

$$\frac{V_o}{V_1} = \frac{-Ap_1}{1 + Ap_o} \tag{16.9}$$

and for node 2 we have

$$\frac{V_o}{V_2} = \frac{Ap_2}{1 + Ap_o} \tag{16.10}$$

where

$$p_o = \frac{1/R_2}{1/R_1 + 1/R_2} = \frac{R_1}{R_1 + R_2} \tag{16.11}$$

$$p_1 = \frac{1/R_1}{1/R_1 + 1/R_2} = \frac{R_2}{R_1 + R_2} \tag{16.12}$$

$$p_2 = \frac{1/R_a}{1/R_a + 1/R_b} = \frac{R_b}{R_a + R_b} \tag{16.13}$$

So the output voltage is

$$V_o = \frac{A(p_2V_2 - p_1V_1)}{1 + Ap_o} \tag{16.14}$$

In the limit $A \to \infty$ we have

$$V_o = \frac{p_2V_2 - p_1V_1}{p_o} \tag{16.15}$$

If we want the output to be proportional to the difference $(V_2 - V_1)$ we must have $p_1 = p_2$ or $R_1/R_2 = R_a/R_b$. The expression for the output voltage then becomes

$$V_o = \frac{R_2}{R_1}(V_2 - V_1) \tag{16.16}$$

Integrator

An op amp can be used to make an almost perfect integrator. This is a circuit where the output is the integral of the input. The transfer function must therefore be proportional to $1/s$. Figure 16.8 shows the op amp integrator.

The graph of the circuit is the same as for the inverting amplifier in figure 16.2. The simplified graph, found by eliminating the "-" node, is also the same as in figure 16.3 for the inverting amplifier. The transfer function is therefore also the same and is given by equation 16.2. The only difference is in the probabilities p_o and p_i. They are given by

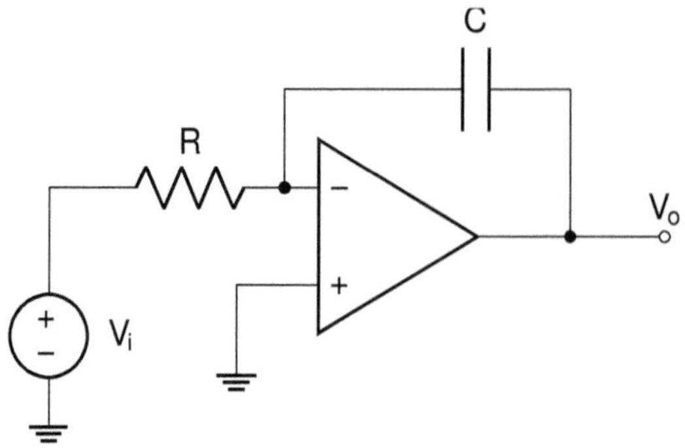

Figure 16.8: Op amp integrator.

$$p_i = \frac{1/R}{Cs + 1/R} = \frac{1}{RCs + 1} \tag{16.17}$$

$$p_o = \frac{Cs}{Cs + 1/R} = \frac{RCs}{RCs + 1} \tag{16.18}$$

The transfer function is then

$$g_o = \frac{-A}{(A+1)RCs + 1} \tag{16.19}$$

For an ideal op amp we take the limit $A \to \infty$ so that the transfer function becomes $g_o = -1/RCs$. The output is therefore the negative integral of the input scaled by the factor of $1/RC$.

17

ACTIVE FILTERS

Low, high and band pass filters using one op amp are shown in figure 17.1. These are known as Sallen-Key filters. Note that all the filters have the same topology and therefore the same graph model. So we only need to calculate the transfer function once using the graph model then we can substitute the particular probability functions for the different filters.

The general topology of a Sallen-Key filter is shown in figure 17.2, and its graph is shown in figure 17.3.

To find the transfer function, start by eliminating the "-" node in the graph. The graph thus reduces to figure 17.4. Next eliminate the "+" node in the graph, leaving the graph of figure 17.5. Finally, eliminate node 1 to leave the graph of figure 17.6. From this final graph we can read off the transfer function. We have

$$g_0 = \frac{V_o}{V_i} = \frac{\frac{Ap_{+1}p_{1i}}{1-p_{1+}p_{+1}}}{1 + Ap_{-o} - \frac{Ap_{+1}p_{1o}}{1-p_{1+}p_{+1}}} \tag{17.1}$$

which reduces to

$$g_0 = \frac{Ap_{+1}p_{1i}}{(1 + Ap_{-o})(1 - p_{1+}p_{+1}) - Ap_{+1}p_{1o}} \tag{17.2}$$

In the limit as $A \to \infty$ we have

$$g_0 = \frac{p_{+1}p_{1i}}{p_{-o}(1 - p_{1+}p_{+1}) - p_{+1}p_{1o}} \tag{17.3}$$

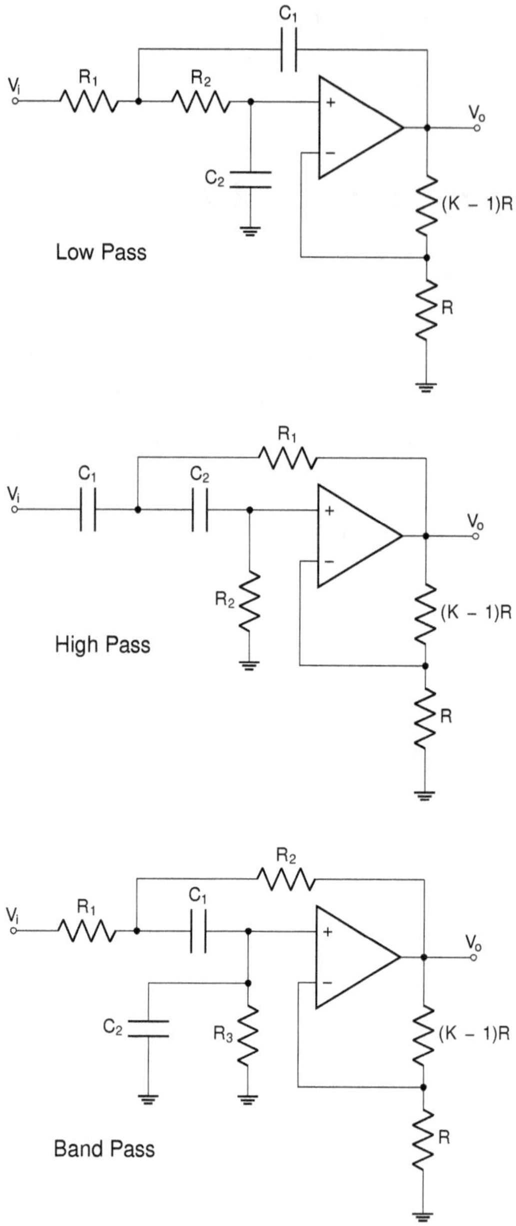

Figure 17.1: Low, high and band pass Sallen-Key filters.

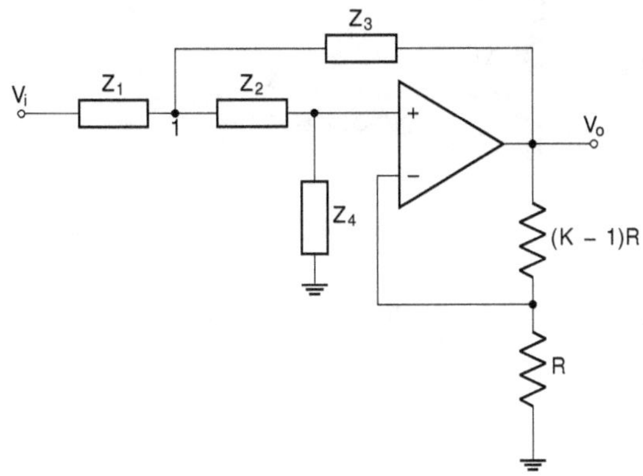

Figure 17.2: General topology of Sallen-Key filters.

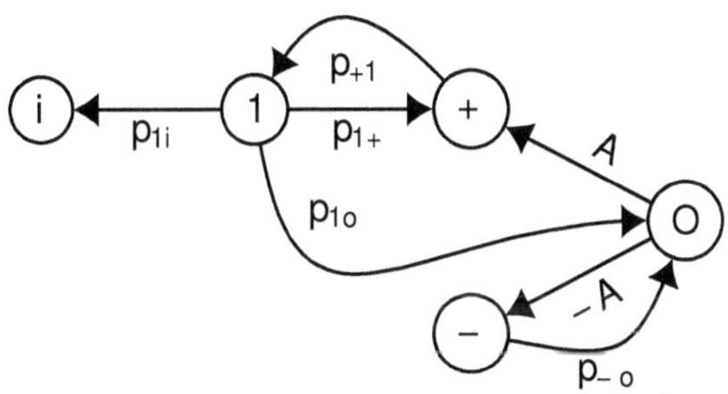

Figure 17.3: Graph of the general topology Sallen-Key filter circuit of figure 17.2.

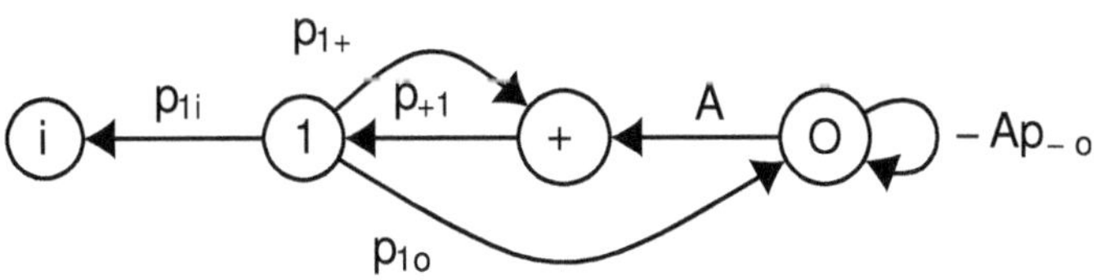

Figure 17.4: Eliminating the "-" node in the graph of figure 17.3.

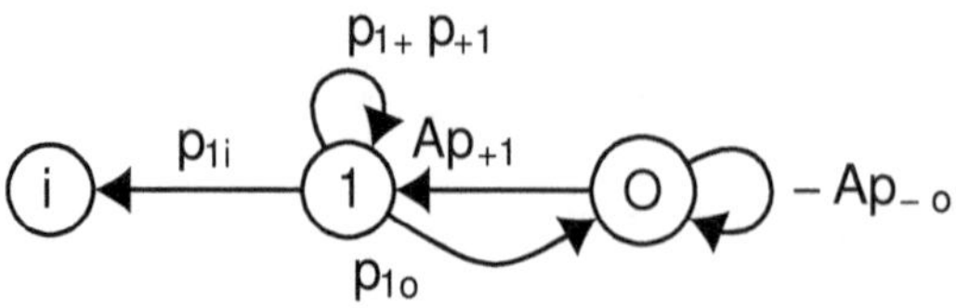

Figure 17.5: Eliminating the "+" node in the graph of figure 17.4.

Figure 17.6: Eliminating node 1 in the graph of figure 17.5.

p_{-o} will have the same value for all Sallen-Key filters, which is

$$p_{-o} = \frac{1/(K-1)R}{1/(K-1)R + 1/R} = \frac{1}{1+K-1} = \frac{1}{K} \tag{17.4}$$

So substituting this into equation 17.3 we get

$$g_0 = \frac{Kp_{+1}p_{1i}}{1 - p_{+1}(p_{1+} + Kp_{1o})} \tag{17.5}$$

Now to get the low pass, high pass and band pass transfer functions we just have to construct the probabilities and substitute them into this equation.

For the low pass Sallen-Key filter, the probability functions are

$$p_{1i} = \frac{1/R_1}{1/R_1 + 1/R_2 + C_1 s} = \frac{R_2}{R_1R_2C_1s + R_1 + R_2} \tag{17.6}$$

$$p_{1+} = \frac{1/R_2}{1/R_1 + 1/R_2 + C_1 s} = \frac{R_1}{R_1R_2C_1s + R_1 + R_2} \tag{17.7}$$

$$p_{1o} = \frac{C_1 s}{1/R_1 + 1/R_2 + C_1 s} = \frac{R_1R_2C_1s}{R_1R_2C_1s + R_1 + R_2} \tag{17.8}$$

$$p_{+1} = \frac{1/R_2}{1/R_2 + C_2 s} = \frac{1}{R_2 C_2 s + 1} \tag{17.9}$$

Substituting these into equation 17.5 we get

$$g_0 = \frac{1}{R_1 R_2 C_1 C_2 s^2 + (R_1 C_1 + R_2 C_2 + R_1(C_2 - K C_1))s + 1} \tag{17.10}$$

To reduce the number of degrees of freedom we can set $R_1 = R_2 = R$ and $C_1 = C_2 = C$. g_0 then simplifies to

$$g_0 = \frac{1}{R^2 C^2 s^2 + (3 - K)RCs + 1} \tag{17.11}$$

For the high pass Sallen-Key filter, the probability functions are

$$p_{1i} = \frac{C_1 s}{C_1 s + C_2 s + 1/R_1} = \frac{R_1 C_1 s}{R_1 C_1 s + R_1 C_2 s + 1} \tag{17.12}$$

$$p_{1+} = \frac{C_2 s}{C_1 s + C_2 s + 1/R_1} = \frac{R_1 C_2 s}{R_1 C_1 s + R_1 C_2 s + 1} \tag{17.13}$$

$$p_{1o} = \frac{1/R_1}{C_1 s + C_2 s + 1/R_1} = \frac{1}{R_1 C_1 s + R_1 C_2 s + 1} \tag{17.14}$$

$$p_{+1} = \frac{C_2 s}{C_2 s + 1/R_2} = \frac{R_2 C_2 s}{R_2 C_2 s + 1} \tag{17.15}$$

Substituting these into equation 17.5 we get

$$g_0 = \frac{K R_1 R_2 C_1 C_2 s^2}{R_1 R_2 C_1 C_2 s^2 + ((1 - K)R_2 C_2 + (C_1 + C_2)R_1)s + 1} \tag{17.16}$$

To reduce the number of degrees of freedom we can set $R_1 = R_2 = R$ and $C_1 = C_2 = C$. g_0 then simplifies to

$$g_0 = \frac{KR^2C^2s^2}{R^2C^2s^2 + (3-K)RCs + 1} \tag{17.17}$$

For the band pass Sallen-Key filter the probability functions are

$$p_{1i} = \frac{1/R_1}{C_1s + 1/R_1 + 1/R_2} = \frac{R_2}{R_1R_2C_1s + R_1 + R_2} \tag{17.18}$$

$$p_{1+} = \frac{C_1s}{C_1s + 1/R_1 + 1/R_2} = \frac{R_1R_2C_1s}{R_1R_2C_1s + R_1 + R_2} \tag{17.19}$$

$$p_{1o} = \frac{1/R_2}{C_1s + 1/R_1 + 1/R_2} = \frac{R_1}{R_1R_2C_1s + R_1 + R_2} \tag{17.20}$$

$$p_{+1} = \frac{C_1s}{C_1s + C_2s + 1/R_3} = \frac{R_3C_1s}{R_3C_1s + R_3C_2s + 1} \tag{17.21}$$

Substituting these into equation 17.5 we get

$$g_0 = \frac{KR_2R_3C_1s}{R_1R_2R_3C_1C_2s^2 + (R_3(R_2(C_1+C_2) + R_1(C_2 + C_1(1-K))) + R_1R_2C_1)s + R_1 + R_2} \tag{17.22}$$

To reduce the number of degrees of freedom we can set $R_1 = R_2 = R_3 = R$ and $C_1 = C_2 = C$. g_0 then simplifies to

$$g_0 = \frac{KRCs}{R^2C^2s^2 + (5-K)RCs + 2} \tag{17.23}$$

18

NEGATIVE IMPEDANCE CONVERTER

Given an impedance Z we want a circuit whose input impedance is $-Z$. One such circuit is shown in figure 18.1. Its graph and its simplification is

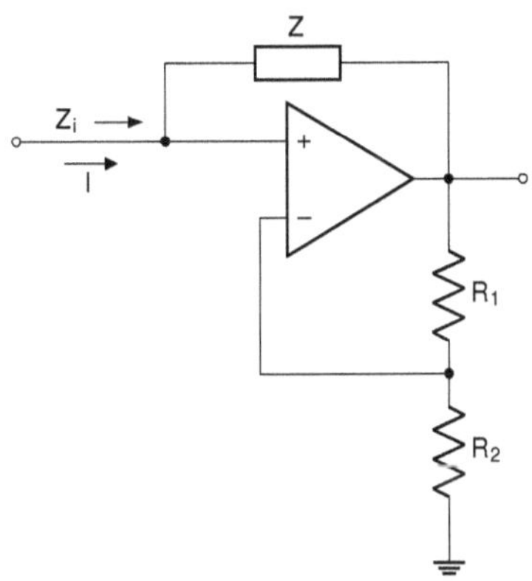

Figure 18.1: Negative impedance converter.

shown in figure 18.2. To find the input impedance we can drive the circuit with a current source and use equation 7.12. The B_{00} in that equation is, in this case, the probability of a walk starting at node "+" and ending at node "+". In simplifying the graph, this means that we want to end with only a loop back probability on the "+" node. This simplification process is shown in figure 18.2. First the negative node is eliminated, then the O node is eliminated until we are left with only the "+" node with a loop

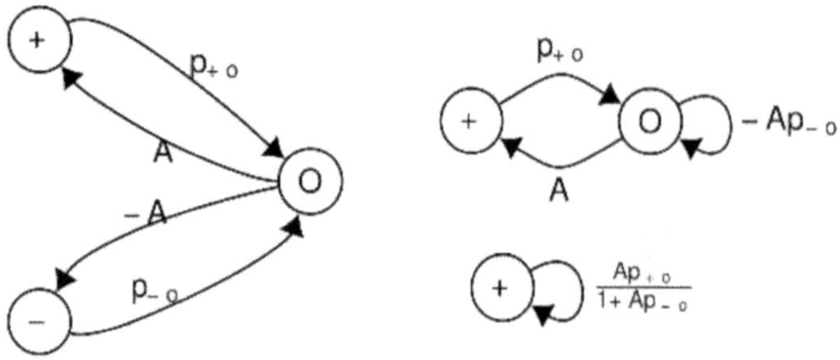

Figure 18.2: Graph and simplification of the negative impedance converter circuit of figure 18.1.

back probability as shown in the figure. So we therefore have

$$B_{00} = \frac{1}{1 - \dfrac{p_{+o}A}{1 + Ap_{-o}}} = \frac{1 + Ap_{-o}}{1 + A(p_{-o} - p_{+o})} \tag{18.1}$$

In the limit as $A \to \infty$ this becomes

$$B_{00} = \frac{p_{-o}}{p_{-o} - p_{+o}} \tag{18.2}$$

The probabilities are $p_{+o} = 1$ and $p_{-o} = R_2/(R_1 + R_2)$. So we get $B_{00} = -R_2/R_1$. The input impedance is therefore

$$Z_i = ZB_{00} = -Z\frac{R_2}{R_1} \tag{18.3}$$

By setting $R_1 = R_2$ we get simply $Z_i = -Z$.

Note that the expression for Z_i in equation 18.3 is symmetric with respect to the Z and R_2 variables. This indicates that we could possibly swap the Z and R_2 elements in the circuit without changing the input impedance. This is indeed the case since doing so does not change the circuit topology. It only changes the transition probabilities, while keeping the same elements. So another way to realize a negative impedance converter is to swap Z and R_2 in figure 18.1.

19

GYRATOR

Passive electrical networks are usually constructed using only four types of elements: capacitors, inductors, resistors and transformers. An engineer is faced with the problem of using these elements to construct a network with specific properties.

But are there really only four elements to choose from? Might there be another overlooked element that will make the network construction problem simpler and more efficient?

These questions occurred to the Dutch engineer Bernard Tellegen while working at the Philips Research Laboratories in the Netherlands in 1948. He determined that there is another possible element which he called the gyrator.

It is a two port device where the current entering a port is determined by the voltage at the other port. The schematic symbol for the device is shown in figure 19.1. The current-voltage relationship between the ports is given by

$$V_2 = RI_1 \tag{19.1}$$

$$V_1 = -RI_2 \tag{19.2}$$

The parameter R is called the gyration resistance. It has the usual dimensions of resistance, i.e. ohms.

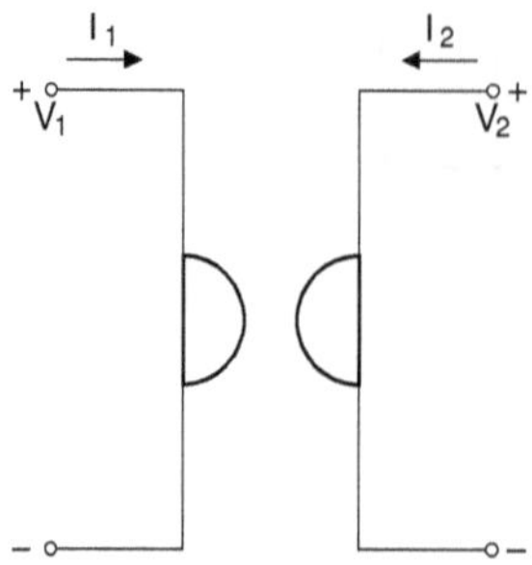

Figure 19.1: Gyrator schematic symbol.

The gyrator is useful not only for designing electrical networks. It also makes it possible to simulate any mechanical system with an electrical network. A mass in a mechanical system is modeled with an inductor. A spring is modeled with a capacitor. Friction is modeled with a resistor. A lever or gear is modeled with a transformer. A gyroscope can be modeled with a gyrator. These five electrical elements can model any mechanical system.

In terms of electrical network design, one of the most useful properties of a gyrator is its ability to invert an impedance. This makes it possible to turn a capacitor into an inductor, or an inductor into a capacitor. Using the current-voltage relationship between the two ports we can find the impedance looking into port 1 as follows

$$Z_1 = \frac{V_1}{I_1} = \frac{-RI_2}{V_2/R} = \frac{R^2}{-V_2/I_2} = \frac{R^2}{Z_2} \tag{19.3}$$

So the impedance looking into port 1 is equal to the inverse of the impedance connected to port 2 multiplied by R^2. If Z_2 is a capacitor then the impedance looking into port 1 will be R^2Cs which is an inductor with inductance equal to R^2C.

It is unfortunately not possible to buy a gyrator in the same way you might buy a capactor or inductor. In other words, there are no passive realizations of a gyrator. But it is possible to make one using two op amps in a circuit called a general impedance converter. The circuit is shown in figure 19.2.

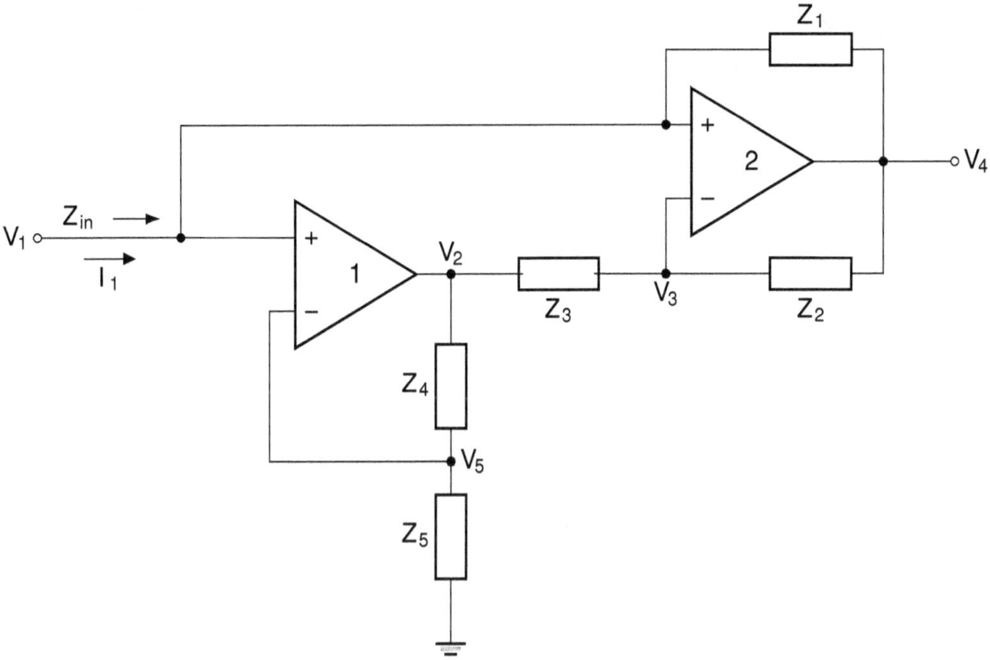

Figure 19.2: General impedance converter.

The input impedance looking into the positive input of the first op amp is

$$Z_{in} = \frac{V_1}{I_1} = \frac{V_1}{(V_1 - V_4)/Z_1} = \frac{Z_1}{1 - V_4/V_1} \tag{19.4}$$

So to calculate the input impedance we need the transfer function V_4/V_1. If we assume ideal op amps (infinite open loop gain) then calculating the transfer function is fairly straightforward. We have

$$V_1 = V_5 = \frac{V_2 Z_5}{Z_4 + Z_5} \tag{19.5}$$

so that $V_2 = V_1(1+Z_4/Z_5)$. The voltage at the negative input of the second op amp is

$$V_3 = V_1 = \frac{V_2 Z_2 + V_4 Z_3}{Z_2 + Z_3} \tag{19.6}$$

Using the expression for V_2 in terms of V_1 given above we can solve the last equation to get

$$\frac{V_4}{V_1} = 1 - \frac{Z_2 Z_4}{Z_3 Z_5} \tag{19.7}$$

The input impedance is then

$$Z_{in} = \frac{Z_1 Z_3 Z_5}{Z_2 Z_4} \tag{19.8}$$

To turn this into a gyrator that converts a capacitor into an inductor, let $Z_4 = 1/Cs$, $Z_1 = Z_2 = Z_3 = Z_5 = R$ then we have $Z_{in} = R^2 Cs$.

If we want to use a one pole rolloff model for the op amps then the analysis becomes more difficult and the best way to do it is to use the random walk method. The graph for the circuit is shown in figure 19.3.

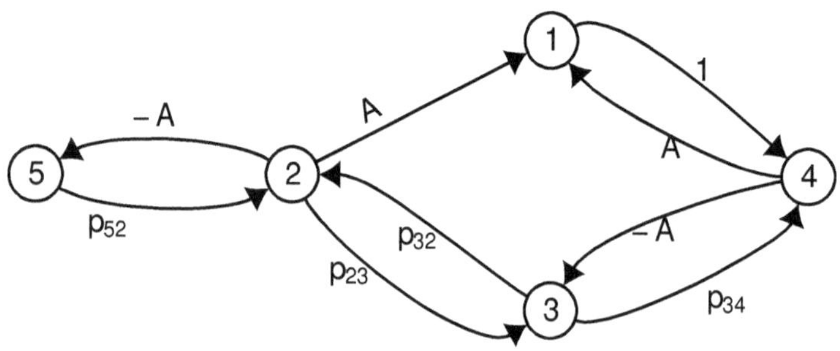

Figure 19.3: Graph for general impedance converter circuit in figure 19.2.

Eliminating node 5 from the graph in figure 19.3 gives figure 19.4.

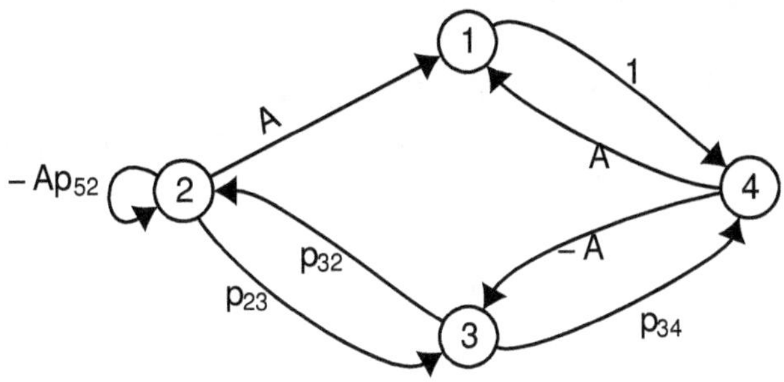

Figure 19.4: Graph in figure 19.3 with node 5 eliminated.

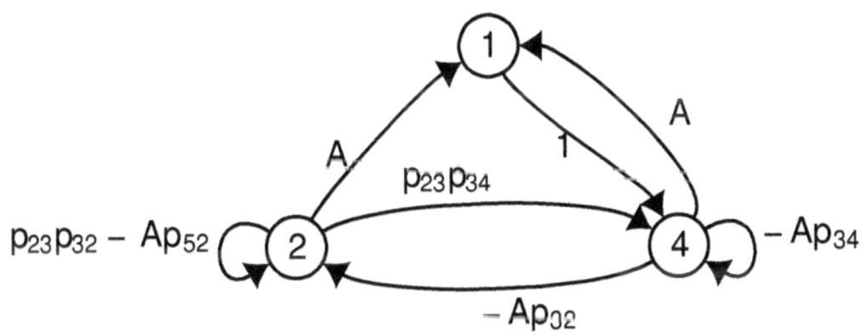

Figure 19.5: Graph in figure 19.4 with node 3 eliminated.

Eliminating node 3 from the graph in figure 19.4 gives figure 19.5. At this point we could continue eliminating nodes or we could use the software discussed in the software chapter at the end of the book to calculate the B_{00} term in equation 7.12. Doing so we get

$$B_{00} = \frac{p_{34}p_{52}A^2 + (p_{34}+p_{52})A - p_{23}p_{32} + 1}{(p_{32} - p_{52} + p_{34}p_{52})A^2 + (p_{34} + p_{52} + p_{23}p_{32} - 1)A - p_{23}p_{32} + 1} \tag{19.9}$$

The only admittance connected to node 1 is $1/Z_1$ therefore the input impedance is

$$Z_{in} = Z_1 \frac{p_{34}p_{52}A^2 + (p_{34}+p_{52})A - p_{23}p_{32} + 1}{(p_{32} - p_{52} + p_{34}p_{52})A^2 + (p_{34} + p_{52} + p_{23}p_{32} - 1)A - p_{23}p_{32} + 1} \tag{19.10}$$

In the limit as $A \to \infty$ this becomes

$$Z_{in} = Z_1 \frac{p_{34}p_{52}}{p_{52}(p_{34} - 1) + p_{32}} \tag{19.11}$$

The probabilities are

$$p_{32} = \frac{Z_2}{Z_2 + Z_3} \tag{19.12}$$

$$p_{34} = \frac{Z_3}{Z_2 + Z_3} \tag{19.13}$$

$$p_{52} = \frac{Z_5}{Z_4 + Z_5} \tag{19.14}$$

With these probabilities Z_{in} is the same as calculated previously in equation 19.8.

20

TRANSISTOR CIRCUITS

With the judicious selection of transistor circuit models, the random walk formalism can be applied to transistor circuits. We will look at the simple example of a common emitter amplifier. The circuit model for the amplifier and the corresponding graph are shown in figure 20.1.

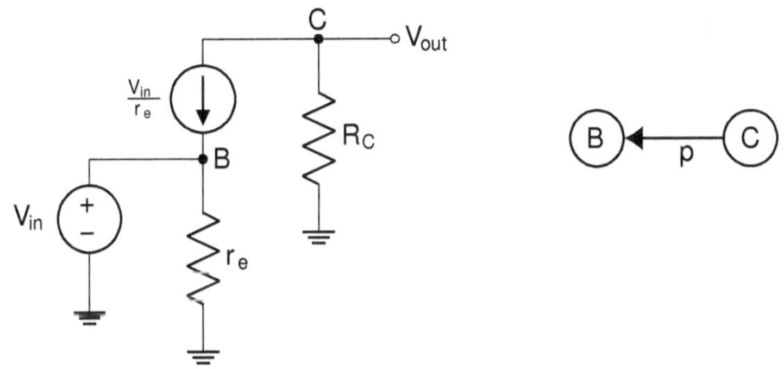

Figure 20.1: Circuit model for a common emitter amplifier and its graph.

This is an Ebers-Moll type model where the collector current is controlled by the base-emitter voltage, which here is just V_{in}. The resistance r_e is called the base-emitter resistance and its value is given by $r_e = V_T/I_C$. I_C is the DC collector current and V_T is a thermal voltage that is related to the temperature by $V_T = kT/q$ where $k = 1.38 \times 10^{-23}$ J/K is Boltzmann's constant, T is the temperature in kelvin, and $q = 1.6 \times 10^{-19}$ C is the electron charge in coulombs. At room temperature $T = 293.16$ K and $V_T = 25.3$ mV.

The value of the collector current source, $V_{BE}/r_e = (V_B - V_E)/r_e$, means that this model allows no current flow into the base of the transistor. This is one of the limitations of this model since it implies that the transistor has an infinite β or current gain and that it has an infinite input resistance.

The graph for this model is as simple as they get. We can write the transfer function immediately as

$$g = \frac{V_{out}}{V_{in}} = p \tag{20.1}$$

To get the transition probability, p, we need the equivalent impedance between the collector and emitter node. That impedance is

$$\frac{V_{out} - V_{in}}{V_{in}/r_e} = r_e\left(\frac{V_{out}}{V_{in}} - 1\right) = r_e(g-1) \tag{20.2}$$

From this we can construct p and hence the transfer function g. We get

$$g = p = \frac{1/r_e(g-1)}{1/r_e(g-1) + 1/R_c} = \frac{R_c}{R_c + r_e(g-1)} \tag{20.3}$$

Now all we have to do is solve this equation for g. We can rewrite the equation as

$$g^2 + (R_c/r_e - 1)g - R_c/r_e = 0 \tag{20.4}$$

This equation can be factored as follows

$$(g-1)(g+R_c/r_e)=0 \tag{20.5}$$

The solution we want is $g = -R_c/r_e$. This is the gain for the common emitter amplifier.

The gain for the common collector can be calculated basically the same way. The circuit model and corresponding graph is shown in figure 20.2.

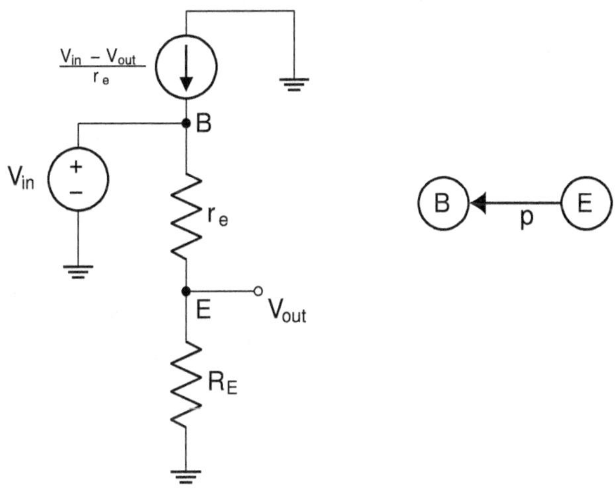

Figure 20.2: Circuit model for a common collector amplifier and its graph.

With the circuit model we can directly construct the transition probability p and hence the transfer function g. We get

$$g = p = \frac{1/r_e}{1/r_e + 1/R_E} = \frac{R_E}{R_E + r_e} \tag{20.6}$$

From this we can see that the gain for the common collector will always be less than 1.

Truth exists for the wise, beauty for the feeling heart.
Friedrich Schiller

21

SOFTWARE

The graphical techniques we used to solve network analysis problems can easily be implemented in software. For example, in the case of nodal analysis with a current source we need only one element of the matrix $B = (I - A)^{-1}$ where A is the transition probability matrix for the network node graph. Equation 7.12 shows that to calculate the equivalent impedance of a network we need only the upper leftmost element of the matrix B. Graphically this is equivalent to eliminating all nodes in the graph except for node 0. The following function written in the Maxima computer algebra system does exactly this given the matrix A. Maxima is a free open source computer algebra system which you can download from https://maxima.sourceforge.io/
The function is called `invert11(A,n)` where A is the matrix and n is the size of the matrix.

```
invert11(A,n):=
block( [i,j,k],
for i:n thru 2 step -1 do
(
  for j:1 thru i-1 step 1 do
  (
    for k:1 thru i-1 step 1 do
    (
      A[j,k]:A[j,k]+A[j,i]*A[i,k]/(1-A[i,i])
    )
```

```
  )
),
return(1/(1-A[1,1]))
);
```

As an example, we will solve for the equivalent resistance of the resistive bridge circuit in figure 9.1. The Maxima code for defining the matrix A is

```
D:a*b+a*c+b*c;
p01:b/(a+b);
p02:a/(a+b);
p10:b*c/D;
p12:a*b/D;
p20:a*c/D;
p21:p12;
A:matrix([0,p01,p02],[p10,0,p12],[p20,p21,0]);
```

If the function `invert11(A,n)` is stored in the file `rwen.mac` and the above definition of A is stored in the file `bridge2.mac` then in a maxima session you would solve for the equivalent resistance as follows.

```
load("rwen.mac");
load("bridge2.mac");
ratsimp(invert11(A,3)/(1/a+1/b));
define(f(a,b,c),%);
```

The function $f(a, b, c)$ defined in the last line will give you the equivalent resistance for any values of a, b, and c. For example, previously we calculated the resistance for $a = 4$, $b = 9$, $c = 6$ so evaluating $f(4, 9, 6)$ gives 6 which is verified by equation 9.12.

For a larger example we will solve for the equivalent resistance of the circuit shown in figure 21.1.

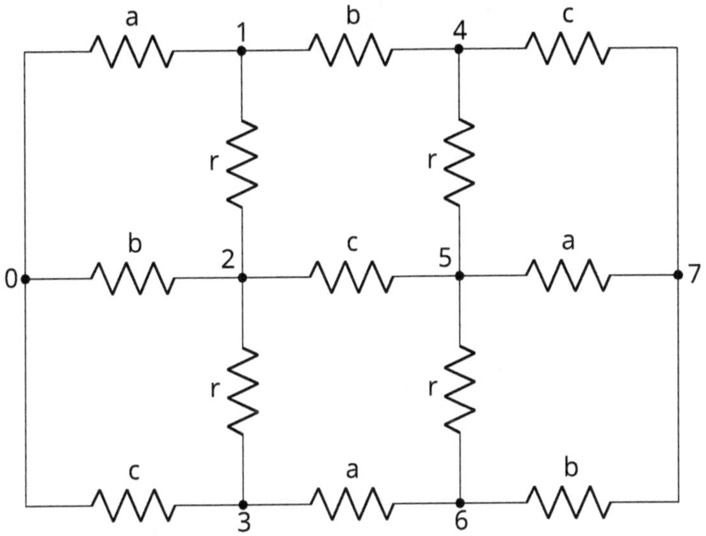

Figure 21.1: A resistive bridge circuit.

The A matrix for this circuit is defined in Maxima as follows

```
D0:a*b+a*c+b*c;
D1:a*b+a*r+b*r;
D2:2*b*c+b*r+c*r;
D3:a*c+a*r+c*r;
D4:b*c+b*r+c*r;
D5:2*a*c+a*r+c*r;
p01:b*c/D0;
p02:a*c/D0;
p03:a*b/D0;
p10:b*r/D1;
p12:a*b/D1;
```

```
p14:a*r/D1;
p20:c*r/D2;
p21:b*c/D2;
p25:b*r/D2;
p30:a*r/D3;
p32:a*c/D3;
p36:c*r/D3;
p41:c*r/D4;
p45:b*c/D4;
p52:a*r/D5;
p54:a*c/D5;
A:matrix([0,p01,p02,p03,0,0,0],[p10,0,p12,0,p14,0,0],
         [p20,p21,0,p21,0,p25,0],[p30,0,p32,0,0,0,p36],
         [0,p41,0,0,0,p45,0],[0,0,p52,0,p54,0,p54],
         [0,0,0,p10,0,p12,0]);
```

If this is defined in the file `bridge3.mac` then we can solve for the equivalent impedance, using $a = 1$, $b = 2$ and $c = 3$, in Maxima as follows.

```
load("rwen.mac");
a:1;
b:2;
c:3;
load("bridge3.mac");
ratsimp(invert11(A,7)/(1/a+1/b+1/c));
define(f(r),%);
```

This gives us $f(r)$ which is the equivalent resistance of the network across node 0 and 7 as a function of r.

$$f(r) = \frac{36\,r^3 + 276\,r^2 + 521\,r + 198}{18\,r^3 + 145\,r^2 + 297\,r + 121} \tag{21.1}$$

The two extremes are at $r = 0$ and $r = \infty$. We get $f(0) = 18/11$ which is 3 times the parallel combination of a, b and c, or $3/(1+1/2+1/3) = 18/11$. We get $f(\infty) = 2$ which is 1/3 the series combination of a, b and c, or $(1+2+3)/3 = 2$. The equivalent resistance therefor ranges between 18/11 and 2 as r goes from being a short circuit to an open circuit.

This is a very basic example of software that can be used to implement some of the methods in this book. The software could be greatly improved by using a sparse matrix representation for A. Other improvements would be calculating the frequency response and writing the software in a general purpose programming language like C. The software used in the above examples can be downloaded from the book's website https://www.abrazol.com/books/rwnet/

Trifles make perfection, and perfection is no trifle.
Michelangelo

REFERENCES & FURTHER READING

- *Network Analysis*, M.E. Van Valkenburg, 3rd ed, 1974.
  https://search.worldcat.org/title/693952
- *Passive and Active Network Analysis and Synthesis*, Aram Budak, 1974.
  https://search.worldcat.org/title/947529
- *Random Walks and Electric Networks*, Peter G. Doyle, J. Laurie Snell, 1984.
  https://search.worldcat.org/title/11764766
- *Finite Automata and Regular Expressions: Problems and Solutions*, Hollos and Hollos, 2013.
  https://www.abrazol.com/books/automata/
- *Combinatorics II Problems and Solutions: Counting Patterns*, Hollos and Hollos, 2016.
  https://www.abrazol.com/books/combinatorics2/

It is not the possession of truth, but the success which attends the seeking after it, that enriches the seeker and brings happiness to him.

Max Planck

ACKNOWLEDGMENTS

In ordinary life we hardly realize that we receive a great deal more than we give, and that it is only with gratitude that life becomes rich. It is very easy to overestimate the importance of our own achievements in comparison with what we owe to others.

Dietrich Bonhoeffer, letter to parents from prison, Sept. 13, 1943

We'd like to thank our parents, Istvan and Anna Hollos, for helping us in many ways.

We thank the makers and maintainers of all the software we've used in the production of this book, including: the Emacs text editor, the LaTex typesetting system, Inkscape, Evince document viewer, Maxima computer algebra system, gcc, bash shell, and the Linux operating system.

That side of our existence whose direction is towards the infinite seeks not wealth, but freedom and joy.

Rabindranath Tagore

ABOUT THE AUTHORS

Stefan Hollos and **J. Richard Hollos** are brothers and business partners at Exstrom Laboratories LLC
(www.exstrom.com) in Longmont, Colorado. They are physicists and electrical engineers by training, and enjoy anything related to math, physics, engineering and computing. In addition, they enjoy creating music and visual art, and being in the great outdoors. They are the authors of the following books:

- **Combinatorics Problems and Solutions, second edition**
- **Creating Noise, second edition**
- **Engineer's Notebook on Inductor and Transformer Circuits: Problems, Solutions and Simulations**
- **The Enigma of the Crookes Radiometer**
- **Passive Butterworth Filter Cookbook**
- **Nell: An SVG Drawing Language**
- **Coin Tossing: The Hydrogen Atom of Probability**
- **Creating Melodies**
- **Hexagonal Tilings and Patterns**
- **Combinatorics II Problems and Solutions: Counting Patterns**
- **Information Theory: A Concise Introduction**

- **Recursive Digital Filters: A Concise Guide**
- **Art of Pi**
- **Creating Noise**
- **Art of the Golden Ratio**
- **Creating Rhythms**
- **Pattern Generation for Computational Art**
- **Finite Automata and Regular Expressions: Problems and Solutions**
- **Probability Problems and Solutions**
- **Combinatorics Problems and Solutions**
- **The Coin Toss: Probabilities and Patterns**
- **Pairs Trading: A Bayesian Example**
- **Simple Trading Strategies That Work**
- **Bet Smart: The Kelly System for Gambling and Investing**
- **Signals from the Subatomic World: How to Build a Proton Precession Magnetometer**

THANK YOU

Thank you for buying this book.

Sign up for the Abrazol Publishing Newsletter and receive news on updates, new books, and special offers. Just go to

https://www.abrazol.com/

and enter your email address.

The web page for this book is also located at abrazol.com.

Joy is the simplest form of gratitude.
Karl Barth

www.ingramcontent.com/pod-product-compliance
Lightning Source LLC
LaVergne TN
LVHW082352200926
848481LV00012B/3542